Integrated Watershed Management

Perspectives and Problems

Integrated Watershed Management

Perspectives and Problems

Edited by

Einar Beheim
Norwegian Water Resources and Energy Directorate
Statens Park, Tonsberg, Norway

Govind S. Rajwar
Department of Botany, Government Post Graduate College
HNB Garhwal University
Rishikesh, Uttarakhand, India

Martin J. Haigh
Department of Anthropology and Geography
Oxford Brookes University
Oxford, UK

Josef Krecek
Department of Hydraulics and Hydrology
Czech Technical University
Prague, Czech Republic

Springer

CP

A C.I.P. Catalogue record for this book is available from the Library of Congress.

ISBN 978-90-481-3768-8 (HB)
ISBN 978-90-481-3769-5 (e-book)

Copublished by Springer,
P.O. Box 17, 3300 AA Dordrecht, The Netherlands
with Capital Publishing Company, New Delhi, India.

Sold and distributed in North, Central and South America by Springer,
233 Spring Street, New York 10013, USA.

In all other countries, except India, sold and distributed by Springer,
Haberstrasse 7, D-69126 Heidelberg, Germany.

In India, sold and distributed by Capital Publishing Company,
7/28, Mahaveer Street, Ansari Road, Daryaganj, New Delhi, 110 002, India.

www.springer.com

Cover illustration: Photograph taken by Martin Haigh in Southern Norway.

Printed on acid-free paper

Printed in India.

Preface

Headwater Control movement believes that headwaters are fragile environments threatened by anthropogenic actions the solutions of which demand practical approach through integrated environmental management. At the same time the Headwater Control calls for direct intervention to mitigate the environmental impacts. Instituted in 1997 as an '*International Association for Headwater Control*', this association promotes sustainable regeneration of headwater regions through practices, awareness, policies and institutions. The approach lies in integrating environmental action such as sustainable management of watershed, various resources and ecosystems in the headwater regions. Sustainable development means not to reduce our habitat's capacity to sustain life and to resolve to leave the land to future generations at least in a condition that was received by us if not better.

The problems of environmental management in highlands and headwaters are the problems of the periphery. Highland and headwater regions, though comprising a nation's economic heartland and often promoted as icons of national identity, are among the least developed and economically backward areas of a nation. These regions consist of great reserves of natural resources which need to be exploited and managed sustainably. The over-exploitation of resources in these regions not only cause a great loss to ethnic and tribal cultures and their traditions, but also sometimes result in a feeling of deprivation of their very resources by the people of the lowland and mainland areas of the country. This often leads to social and political unrest in these areas. In some cases, the resources are shared by adjoining nations which may cause trans-national problems.

The International Conference on Headwater Control held at Bergen, Norway during June 20-23, 2005 was the sixth of its kind (IHC6) belonging to a tradition established at the First International Conference in Prague in 1989 and developed through subsequent meetings at various places in Europe, Asia and Africa. IHC6 discussed various issues concerning headwater regions of the world under wide-ranging themes: climate change impacts, vegetal cover, sub-surface hydrology, catchment and streamflow hydrology, pollution, water quality and limnology, remote sensing and GIS, environmental impact assessment and mitigation, socio-economic impacts, public participation, education and management, and integrated watershed management. Twenty two articles from over 200 oral and poster presentations were selected, updated and peer reviewed for inclusion in this volume as proceedings of the IHC6 conference. The issues of water use and water management in different parts of Europe, Africa and Asia have been the central theme of most of the articles, which are divided into four sections: A. Sustainable watershed management; B. Catchment and streamflow hydrology; C. Quality, pollution and management of water resources; and D. Monitoring and mitigation of disasters.

In the first article, Martin J. Haigh has shed revealing light on the importance of linking themes of water supply and better water management. These works suggest four arbiters for success in the sustainable management of headwaters: a clear and agreed message, clearly targeted action, a willingness to work with others (not least the UN agencies), and personal commitment as the main motivation. Haigh is of the opinion that improving the management of headwater watersheds, as with all environmental management, requires a shift in social attitudes away from granting primacy to short-term economic need towards a more Earth-centred vision.

Water is an essential resource to be tapped for sustaining life and development. Ted L. Napier in his article stresses the need for social scientists to engage in research and programme planning and its implementation in coordination with physical scientists in multiple object decision making within watersheds. He argues the importance of social science involvement in interdisciplinary research and planning to be one of the most critical components of watershed management that can also play a critical role in assessing natural resources development needs and priorities among watershed residents. Such information will make it possible for watershed residents to become more involved in the determination of future natural resources development planning. Increased public involvement in decision making should result in a significant increase in the social acceptability of natural resources development planning and subsequent programme implementation.

In his article on assessing socio-economic and resource sustainability in managing headwater regions in Australia, John Cary considers the analysis of socio-economic sustainability within the framework of an interdisciplinary study synthesising bio-physical, land capacity, ecological, economic and social aspects of land management in three upper river catchment areas. An example of an operational framework for the analysis of the social, economic and ecological resilience of land use and land management in an integrated river system in the Murray-Darling Basin, the largest river basin in Australia, has been presented.

Pier Carlo Zingari, in his article "Building co-operation, coalitions and governance on mountain catchments sustainability", points out the importance of mountains as significant areas for hydrology and ecology providing benefits to large areas upstream and downstream. He highlights the means to build on the knowledge base, the capacities, the exchange of experiences, governance and communication.

Leif Lillehammer, Terje Kleven, Tore Hagen, Mark Bain and David Lewis in their article seek to develop sustainability priorities with a participatory process in the Lake Victoria Basin in east Africa. They explain the Shared Regional Vision and its various follow up actions as a tool for reducing conflict between the Lake Victoria states themselves over their joint recourses, also leaving them in a more joint and harmonized position while discussing about utilization of the Lake Victoria headwater resources with the downstream Nile states. The Regional Shared Vision of Lake Victoria basin and the NBI Shared Vision provide a platform for further collaboration and dialogue related to utilization of resources of the Nile headwater region.

Atle Hindar, Yvan Orsolini and Brit Lisa Skjelkvåle in their article discuss the management of headwaters in acidified areas along the west coast of Norway. Watercourses along the west-coast of Norway are relatively short, but specific runoff is high and rapidly changing due to high precipitation amounts, steep gradients and thin soils. Large-scale climatic patterns interact with long-range transported air-pollutants to reduce the water quality in general and in episodic events. Watercourses close to the coast are impacted by high deposition of sea-salts, especially during extreme weather conditions. Sea-salts may be ion-exchanged with aluminium and hydrogen ions in the acidified soils of the area and thus influencing the quality of surface waters. The resulting impacts on Atlantic salmon populations and invertebrate communities have called for measures to improve the water quality. Their work links the various factors and shows water management approaches under the relatively extreme hydrological conditions in this area. Whole-catchment liming may be especially interesting as a long-term mitigation strategy for acidified headwaters.

Ian Cordery and Peter S. Cloke have stressed the need for environmental and hydrological monitoring and modelling in their article. As little hydrological monitoring is occurring in the world today, they emphasise that hydrological monitoring has repeatedly been shown to provide economic benefits which are far greater than the costs of observing, editing and archiving the resulting data. They state that scientists and engineers who understand the problem need to explain it to their communities, politicians and CEOs in a clear, unmistakably persuasive manner.

In another article on headwater region of Nile in Ethiopia, Adrian Wood, Ato A. Hailu and Alan Dixon share different type of degradation responsible for wetland loss. The Wetland Action applied in this area has provided technical inputs into community management of wetlands in the Nile Headwaters in Rwanda and in Zambezi headwater wetlands in Zambia. They have suggested specific sustainable wetland use and rehabilitation for its restoration. They conclude that attempts in Ethiopia towards solving these problems is generating lessons for use in several other wetland areas of Ethiopia and also in other parts of the Nile Basin and beyond.

J. Wilk, L. Andersson, P. Wolski, D. Kgathi, S. Ringrose and C. Vanderpost have described the upstream development and downstream effects of changing flow in the Okavango Basin. They indicated that the impacts of development scenarios are considerably smaller than the predicted impact of climate change. The simulated hydrological effects show changes in the Delta ecosystems; in the area of permanent swamp, seasonal floodplains, and intermittently flooded areas. The combined effects of human abstraction and climate change may result in significant changes to Delta flooding regimes with significant ecological and human effects.

Atul H. Haria and Paul Shand in their article have investigated the mechanisms of upland streamflow generation by a detailed physico-chemical study in the forested Plynlimon catchment (headwater of the Afon Hafren-River Severn) in mid-Wales, UK. Groundwater levels in the depth-specific boreholes indicated that the deeper groundwaters were confined or semi-confined at the study site close to the river; however, all groundwaters responded rapidly to rainfall. Borehole geophysics confirmed flows located in discrete fractures in the bedrock. Chemical stratification of waters sampled through the boreholes further defined the depth-specific nature of these groundwaters. Additional groundwater analyses showed mixing between shallower (younger) and deeper (older) waters.

In their article, Marco Cantonati, Ermanno Bertuzzi and Alessia Scalfi make biodiversity assessment and integrity evaluation of springs of Trentino and long term ecological research in Italian Alps incorporating various measures of hydrologic, hydrochemical and biological data in the preliminary investigations. The Headwater Control Conferences address the issues concerning the sustainable management of water and other natural resources so that the human life can be sustained and the resources can be saved for future generations too.

Helena Nynas highlights the importance of water as an element in building the nation of Norway and the Norwegian identity, where 389 rivers are permanently protected from hydropower development and four of the 12 highest waterfalls are protected. Taking into account the considerable pressure for hydropower development Nynas presents relatively great success achieved by Norway in nature conservation through Norwegian way of feeling and using nature on a daily basis by providing us with the fact that almost a fourth of the Norwegian hydropower potential is still protected.

David A. Wright, Kjetil Sandsbråten and Olav Osvoll in their article present an assessment of water resources and water use in Angola using modern remote sensing techniques combined with available data and geographical information system (GIS) to make national and regional overviews of rainfall, flow, groundwater, evaporation, population, water use and other parameters. In Angola decades of civil war and uncharted landmine fields played their toll on the country's infrastructure and its ability to utilise these resources with much of the potentially productive areas lying fallow. The Government of Norway has sponsored a programme of support to the water sector in Angola including institutional cooperation between the National Directorate of Water (DNA) and the Norwegian Water Resources and Energy Directorate (NVE).

The downstream perspective in water management in middle mountain catchments of the Nepal Himalayas has been discussed by a team of workers led by Juerg Merz. In this study, the area of cultivated land showed a negative correlation with the flood peaks and volumes at the sub-catchment outlets and grassland and degraded land showed a positive, while the forest areas did not show any distinct correlation. In order to understand upstream-downstream sediment linkages, it is important to consider interventions to reduce the sediment loads from these middle mountain catchments for downstream development. The floods at the different sub-catchment outlets further show very high correlation with the processes on the degraded and grassland plots. This suggests that factors like abandoning of cultivated areas and outmigration increase the flood generation potential. While the importance of catchment characteristics is proven for small to medium flood events, at high flood events all land use types are contributing to these floods in the same manner and only rainfall characteristics are decisive.

Preeti Saxena and Asim K. Bhattacharyya have evaluated the environmental risk associated with hazardous waste generated in small scale industrial areas of Delhi. Contamination by heavy metals, organic nutrients, toxic wastes and microbes in industrial areas have caused serious hazards to land and water.

Marina Pintar and co-workers in their article describe the impact of land use on nutrient concentration in upper streams of waters in Slovenia. The main purpose of the work was to examine if there was any statistical relationship between land use and nitrogen and phosphorus concentration in upper part of 35 Slovene water streams. The land use data were combined with the data of nitrate, ammonium and phosphorus concentration in surface water. Stepwise regression procedure was used in the selection of independent variables for the models. They observed significant statistical relationship between the shares of different land use in the watershed and ammonium, phosphorus and nitrate concentration in surface water on sampling station at the 99% confidence level.

J. Krečec and Z. Horicka in their article report the recovery of headwater catchments and lakes affected by the acid atmospheric deposition. In the 1990s, the observed recovery of surface waters (an average increase of pH values from 4.0 to 5.0, and drop in concentrations of aluminium from 1.2 to 0.4 mgl^{-1}, successful reintroduction of brook char) results particularly from the decreased air pollution, reduced leaf area by the harvest of forests, and liming of both reservoirs and watersheds.

Mandira S. Shrestha and Karma S. Chhophel describe disasters in the Hind Kush Himalayan region citing a case study of the Tsatichhu Lake of Bhutan. After the occurrence of landslides, an early warning system was developed for preventive measures. The case study shows the early wireless warning system established after occurrence of landslide. Dams had successfully reduced the loss of property and prevention of loss of life in downstream. Shrestha and Chhophel recommend building community awareness, mapping of environmental hazards and use of early warning system as essential in disaster reduction in mountain region.

Irina Krasovskaia, Anja Skiple Ibrekk, Lars Gottschalk and Hallvard Berg in their article show how information may be used for streamlining knowledge on flooding. The views of decision-makers on flood hazard were studied by means of expert panel discussions at national and international levels. Both poll and focus groups studies revealed an underestimation of flood risk and passive attitudes towards flood problems. Within the project a number of different disseminating techniques have been demonstrated which include school projects, awareness campaigns, information desks, leaflets to inhabitants, improved flood warning and new ways of visualising flood risk. They suggest the results of FLOWS (Flood plain Land use Optimising Workable Sustainability) project as an important measure in all aspects of flood management with special emphasis on spatial planning.

Roelof J. Stuurman, Perry G.B. de Louw and Marc F.P. Bierkens have tried to decrease the risk of floods in small and medium-sized catchments through natural storage in headwater and riparian zones taking a case study of Merkske catchment of Europe. They conclude that in the Brook valley groundwater-oriented level management may be useful at some places.

In a field study on sediment mobilisation, Martin J. Haigh has shown a direct relationship between soils and sub-soils that suffer erosion, and the sediments that are deposited as the result of erosion in different sediment traps. He concludes that the sediment that is deposited in the short term may not represent the source materials directly but originate from the bed and wash load of a channel or landslide deposits sourced from a channel bank.

In the concluding chapter, Seifeldin H. Abdalla states that the Blue Nile and its tributaries and Atbara river, originating from the Ethiopian plateau, are the main source of sediment in the canalization system, dams' reservoirs, and inlet channels of the pumping schemes. The economic and environmental impacts have been evaluated in terms of loss of fertile agricultural lands, crops and high value perennial trees and dislocation of families and villages.

These selected articles from amongst several presentations at Bergen conference constitute a broad spectrum of issues of integrated headwater management in Asia, Africa and Europe. Wide-ranging issues discussed by various experts deal with water quality and pollution, water management, sedimentation, monitoring and mitigation of disasters, soil erosion and conservation, sustainable management and utilisation of headwater resources, participatory management practices, socio-economic studies, impacts of land use change, biodiversity assessment and recovery processes in Himalaya and other highland regions of different parts of the world. The book would serve as a tool and reference for various watershed management practices for planners, environmental managers, water policy makers, ecologists, biologists, geographers, geologists, social scientists and climatologists.

Einar Beheim
Govind S. Rajwar
Martin J. Haigh
Josef Krečék

About the Editors

Dr. Einar Beheim, a founding member and National Representative of the International Association for Headwater Control, has recently retired as Regional Manager for NVE, the Norwegian Water Resources and Energy Administration. Einar Beheim was the lead organiser and host for "*Headwaters 95: Hydrology, Ecology and Water Resources*", the Sixth International Conference on Headwater Control, which was held in Bergen, Norway, 20-23 June 2005. The conference was co-sponsored by Norwegian Water Resources and Energy Administration (NVE) and International Association for Headwater Control with support from the International Association for Hydrological Sciences, World Association of Soil and Water Conservation, United Nations University, UNESCO, International Union of Forest Research Organisations, European Observatory of Mountain Forests, the University of Bergen and City of Bergen, Olje-Og Energie department.

Dr. Govind S. Rajwar, working as Associate Professor of Botany, Government Post Graduate College, Rishikesh, India, has been elected as a Fellow of the Linnean Society of London (2007) and of the National Institute of Ecology (2008), and a Life Member of the National Academy of Sciences India (2006). He conducts research on forest ecology, biodiversity, environmental impact studies and conservation of the Himalayan region. He has been editor/associate editor of the journals *Vegetos* and *Environmental Research Journal* and reviewer for many international journals including *Journal of Environmental Management, International Journal of Biodiversity & Conservation, Journal of Horticulture & Forestry, Eurasian Journal of Biosciences, Carbon Balance & Management* and *Journal of Herbs, Spices & Medicinal Plants.* He has published/edited six books on ecology of Himalaya and headwater management and 85 research papers in various national and international journals. He has supervised research work on many projects funded by various agencies. He has received some awards and recognitions including Brilliant Young Scientist Recognition by the XIV International Congress, Germany (1987) and Excellence Award for outstanding contribution to ecology and environmental awareness (1998).

Professor Martin J. Haigh is a Co-Director of the International Association for Headwater Control, a former Vice-President of the World Association for Soil and Water Conservation, and until December 2009, Co-Editor of the *Journal of Geography in Higher Education*. He is on the Editorial Board of *Asian Journal of Water, Environment and Pollution* since 2004. He is, currently, Professor of Geography and University Teaching Fellow at Oxford Brookes University in England, and Senior Fellow of the Higher Education Academy of the UK. He conducts research into Education for Sustainable Development and Community-based Environmental Reconstruction.

Dr Josef Krečeк is the founder and managing Co-Director of the International Association for Headwater Control, a prize-winning Principal Investigator for the international NGO 'Earthwatch', and was feted as a 'European Champion' (Environment) by *Time* magazine. He teaches Applied Hydrology at the Czech Technical University in Prague, Czech Republic and conducts forest hydrological research in the mountains of Bohemia and Slovakia.

Contents

SECTION A

Sustainable Watershed Management

1

Sustainable Management of Headwater Resources

Martin J. Haigh
Department of Anthropology and Geography
Oxford Brookes University, Oxford, UK

Introduction

The 6th International Conference on Headwater Control (IHC6) was held in Bergen in 2005. This meeting on Headwaters belongs to a tradition established at the first international conference in Prague, 1989 (Krecek et al., 1989) and developed through subsequent meetings (Krecek and Haigh, 1992; Singh and Haigh, 1995; Haigh et al., 1998). Much of any progress made is built upon the foundations of these predecessors and, hopefully, will provide the building blocks from which future headwater understanding will be constructed. IHC6's immediate antecedent was IHC5—'*The International Conference on Sustainable Management of Headwater Resources*'. This was organised in Nairobi in September 2002 (Jansky et al., 2006; Haigh, 2004). The aim of this extended report is to provide a record for those who did not make the journey to Nairobi and an *aide memoire* for those who did (Jansky et al., 2006).

The context for the 2002 meeting was WSSD, the World Summit on Sustainable Development in Johannesburg. However, this year also has its own special significance. In January 2005, the '*United Nations Decade of Education for Sustainable Development*' was launched. Education for Sustainable Development is defined in many ways but UNESCO, the UN's lead agency, talks about "*learning how to make decisions that consider the long-term future of the economy, ecology and equity of all communities*" (UNESCO, 2003, p. 4). Sustainable development means learning to live on the Earth as if the future mattered. It means not reducing our habitat's capacity to sustain life and resolving to leave the land to future generations in a condition that is as at least as good as it was when it was received by our own.

People rarely intend to create environmental problems. They result *"as the sum of a myriad of seemingly miniscule individual actions... [but] motive is irrelevant to environmental impacts"* (Schneider, 1996, p. 6). The challenge of this United Nation's Decade is to prove that we can live non-destructively upon the Earth. The then Secretary General of United Nations, Kofi Annan, has pointed out that: *"Sustainable development will not happen of its own accord. We need a break with the harmful practices of the past..."* (Annan, 2001, p. 2). Unfortunately, these practices are fostered by a culture that always finds 'someone else' to blame for the world's problems – the rich, the poor, big business, bad government, bad weather, bad luck . . .'; the list is interminable. The truth, of course, is rather less palatable. Each of us contains the root cause of unsustainable development. Sustainability is affected by every thing we do and, in this sense, land degradation is caused by us all (Bridges et al., 2001). We are, individually, responsible through each lifestyle and work decision we make. Equally, it is a personal duty to try and mitigate our impact.

Kofi Annan summarises: *"Our biggest challenge in this new century is to take an idea that sounds abstract – sustainable development – and turn it into reality for all the world's people"* (Annan, 2001, p. 2). This is a huge challenge for education. One technique used to convey the message is 'ecological foot printing'. This calculates our share of the Earth's resources and how much more than that share we actually use (Chambers et al., 2001). The technique, while extremely crude, has one important virtue. It expresses the problem of sustainable development in personal terms. The message that it conveys is that 'what I do makes a difference'. Sustainable development is a personal matter and the challenge it poses to every one of us is "what can I do that may make a difference?"

As the headwater and watershed management professionals, we have a special responsibility. We own some of knowledge and skills needed 'to make a difference'. The world may worry far more about other matters, but we know that environmental security and the defence of environmental integrity rank among the most serious threats. Land degradation reduces the productive capacity of existing lands and encourages their inhabitants to move on, perhaps to degrade more habitats elsewhere. Environmental degradation is commonly caused by poor land husbandry, which is exacerbated by factors that often begin with poverty, social exclusion, inequity, conflict, failing social infrastructures, adverse policy environments, and economic globalisation. The problems of headwaters are much more than those of physical environmental control. The 'Cinderella' of headwater research remains its social scientific base. However, the key to effective headwater control is community engagement in, economically and environmentally, sustainable better land husbandry.

Results from 'The International Conference on Sustainable Management of Headwater Resources', Nairobi, Kenya September 2002 (IHCS)

IHC5 conceived itself as a break-out from the United Nations WSSD Conference in Johannesburg. It was cosponsored by several UN agencies, principally, the United Nations University, so, appropriately enough, an international declaration was produced as a major output. The *Nairobi Declaration for the International Year of Freshwaters 2003* was ratified by five United Nations agencies, as well as the International Association for Headwater Control. Its text, which largely reflects the contemporary concerns of WSSD, appears as Box 1 (Jansky et al., 2003).

Box 1: Nairobi "Headwater" Declaration for the International Year of Freshwater 2003 (see: Haigh, 2004; Jansky et al., 2003).

We, the participants in the International Conference on Sustainable Management of Headwater Resources, held in Nairobi (Kenya) on 5-8 September 2002,*

1. Acknowledging with gratitude the United Nations General Assembly Resolution No. 55/196 to declare the year 2003 as the International Year of Freshwater, thus drawing the world's attention to the need to foster sustainable development and management of freshwaters;
2. Noting the outcome of the recent World Summit on Sustainable Development in Johannesburg 2002, where commitments were made "to increase access to clean water and proper sanitation, to increase access to energy services, to improve health conditions and agriculture, particularly in drylands, and to better protect the world's biodiversity and ecosystems";
3. Recognizing that headwater regions are sensitive environments, source areas for both surface and groundwater resources, and lands that affect the quality of freshwater supplies;
4. Keeping in mind that headwater regions lie at the margins of both watersheds and, often, social and economic systems;
5. Recognizing also the critical environmental functions of headwater regions and their importance for the livelihoods of both their inhabitants and for those who inhabit lands downstream as evoked in Chapter 18 *"Protection of the quality and supply of freshwater resources: application of integrated*

* This conference, which took place from 5-8 September 2002 at the United States International University—Africa in Nairobi, Kenya, was jointly organized by the United Nations University, Tokyo, Japan, UNESCO-Nairobi, the United Nations Centre for Human Settlements (UN-HABITAT), the United Nations Environment Programme (UNEP), USIU and Kenyatta University, Nairobi, Kenya, in collaboration with International Association for Headwater Control (IAHC); International Association of Hydrological Sciences (IAHS); and the World Association for Soil and Water Conservation (WASWC).

approaches to the development, management, and use of water resources" of Agenda 21 adapted at the United Nations Conference on Environment and Development (1992), and also as stipulated in its Chapter 13 entitled *"Managing fragile ecosystems: sustainable mountain development";*

6. Affirming our concern to mitigate the consequences of the increasing human impact in headwater regions caused by competing demands for water, forestry, agriculture, energy production, tourism, transport and urban development, which continue to affect the environment adversely, not least with respect to the provision of clean water supplies and the maintenance of other hydrological functions;
7. Noting with further concern that policies can impair, seriously and inadvertently, the course of headwater resources management, and that this can create problems downstream for the quality, quantity and distribution of available freshwater resources;
8. Recognizing that sustainable management of headwater regions needs a holistic, integrated approach which respects the needs of all stakeholders in the regions, values and empowers the headwater inhabitants, and which recognizes their central role in the stewardship of headwater systems;
9. Affirming that the sensitive and scientific management of natural resources, supported by improved access to the high quality data required is essential for fostering development that is not only sustainable, but ideally self sustaining;
10. Conscious that unsustainable management has negative impacts on the health, productivity, social and economic welfare and ecosystems of headwater regions;
11. Aware of the increasing demand for potable waters that will be required for human health, welfare and well-being, and of the crucial role that headwater regions will play in meeting this demand; and
12. Conscious also of the potential negative interactions between the inhabitants of headwater regions and those downstream, caused by competition for the limited resources available in the regions, and aware also that headwater areas accommodate and provide for livelihood of a large number of populations, whose activities and resources consumption may have significant effects on the well-being of those who live downstream;

Declare that:

13. Sustainable development should be the baseline for all environmental policy, planning, management practice, education and law in headwater regions;
14. UN agencies should continue their work with all stakeholders to appraise their situations, to identify gaps in knowledge, needs and constraints, and to support them in their efforts to resolve their problems and undertake practical action towards more self-sustaining and environmentally sensitive development;
15. An 'international commission' for headwater management should be established in order to provide direction and continuity for headwater issues and to create an awareness of headwater concerns at governmental level;

16. Priority should be given to the creation of new management structures at all levels, which should be designed to improve the coordination, cooperation and empowerment of all stakeholders of headwater regions, not least to enhance the participation of women, disadvantaged social groups and minority communities, and to tap and develop the full spectrum of local indigenous knowledge relating to watershed planning and management;
17. Greater effort should be devoted to the refinement of methods for generating and sharing the appropriate and reliable information needed for environmental research, planning and management and also for the transfer of appropriate low cost technologies, especially with respect to 'cushioning' the impacts of environmental hazards for human populations;
18. Greater attention needs to be paid to the special roles and hydrological functions of headwater wetlands and peat lands, which should be a special focus for future headwater workshops, and also to the impacts of anthropogenic processes on watershed functions in headwater regions;
19. The quality of life for the inhabitants of headwater regions should become a primary concern, including the basic needs for a healthy environment and the regeneration of degraded headwater habitats where required;
20. Greater attention should be paid to applied environmental education aimed at building capacity for headwater management and changing social attitudes against wasteful and polluting uses of headwater resources;
21. NGOs (community-based non-governmental organizations devoted to environmental and/or social uplift) should be empowered to play a greater role in the planning, regeneration and management of headwater habitats, by promoting more efficient mechanisms for financial support for effective NGOs;
22. Greater attention should be given to management of headwaters in arid and semi-arid lands, especially with respect to groundwater management and improved accessibility of potable waters to headwater inhabitants, while one of the main focuses should be to reduce the time wasted in carrying water to households from distant water sources;
23. Attention should also be paid to alternative measures that would reduce the dependence of downstream areas on the resources of headwater areas, including reducing wastage and increasing the efficiency of resource utilization, not least of water;
24. The equitable distribution and use of headwater resources remain a major concern, and planning and management of headwater regions needs to be integrated within the broader framework of watershed management that addresses the concerns of both headwater inhabitants and those downstream.

We therefore call upon UNU, UNESCO, UN-HABITAT, FAO, UNEP, UNDP and other concerned international and national organizations, governments of both developed and developing countries, corporations and NGOs, to facilitate headwater research, monitoring, capacity-building, self-sustaining sustainable development, and better management of the headwater environments, and to help create linkages and synergies in this regard among environmental managers, scientists, communities, policy/decision-makers, practitioners and the general public.

Sustainable Management of Headwater Resources

However, the Declaration was by no means the only product of the discussions in Nairobi. This paper aims to remind delegates to its successor conference, IHC6, Bergen, of the other progress made in Nairobi. Here, the discussion addressed were two key dimensions of the sustainable development of water resources in Africa and South Asia. The first were the social processes of sustainable development, especially the efforts being made to transform a worldview that allows land degradation and desertification into something that can help communities live self-sustainably in their habitats. The second concerned science—because, in the words of Stephen Schneider, *"Ignorance of the laws of nature is no excuse"* (Schneider, 1996, p. 6). In the following section, all named contributions are related to papers in the IHC5 Proceedings (Jansky et al., 2006).

Several contributors struggled to understand environmental processes and environmental change. Subramanian evaluated data from the Ganges and Brahmaputra Rivers finding that gross sediment erosion and accumulation rates reflect a combination of present hydrological and long-term tectonic processes rather than any human impact. Several contributors constructed models to evaluate local environmental change. Onyando and Chemelil's worked on small ungauged catchments showing that combining the SCS runoff curve and Nash cascade model for runoff-routing allowed the creation of a useful rainfall-runoff model for flood prediction. Obando's model of the impacts of human migration and scrub clearance in semi-arid Kenya found that increasing aridity need not accelerate erosion. However, Onywere's study of the headwaters of Lake Bogoria, Kenya, a major tourist destination famed for its flamingoes, geysers and scenic beauty, which is also home to the Jemp and Tugen peoples, finds that the area suffers severe land degradation and the lake ecosystem, which supports its tourist industry and biodiversity, is threatened by agricultural pollutants.

However, many expressed concern about the impact of climatic change (IPCC, 1996), which Gathenya called "the most profound global threat facing humanity in this millennium". Indeed, some recent reports suggest these processes are moving twice as fast as previously suspected (SAPA-AFP, 2005). Meanwhile, Gathenya's work evaluated the impact of climate change on Kenya's Thika catchment (area: 1852 km^2), which provides water to Nairobi and several irrigation projects. It found that small changes in precipitation could have dramatic impacts on river flow depths.

In fact, many papers reported major change in the flow of rivers and springs, although few attributed these to climatic changes. Schutt and Thiemann, who studied water levels in a shallow endorheic Ethiopian lake, found no link between climate variability and lake levels. By contrast, flows in a tributary river declined steadily throughout the period. Population growth in its headwaters had encouraged deforestation and scrub clearing, while agricultural intensification led to increased water usage and accelerated erosion.

Similar results came from Murimi and Prasad, who studied Lake Nakuru, another shallow endorheic lake in the central Rift Valley. This highly alkaline lake's depth and area had changed dramatically during the last 100, but not in recent, years. Lake level fluctuations were attributed to climatic and/or geologic factors and/or land use changes, notably the clearance of forest for the resettlement of 30,000 families in a state agricultural colonisation programme. Once again, the team reported a dramatic decline in local river flow magnitudes. Especially marked was an increased frequency of low flow events across all recurrence intervals and all locations except the headwaters. Murimi and Prasad suggested that the changes were related to water abstraction for domestic and irrigation purposes.

In Kenya, water supply is ca. 640 m^3a^{-1} per person. This is projected to decline to 170 m^3 by 2050 through population growth. In the Mount Kenya region, population growth had increased the need for water for drinking, industrial, urban and, most important, irrigation uses. Currently, ten times more water is taken than official regulations permit. Opere's records show that mean annual flow in the Ewaso Ngiro North has decreased from 82.36 m^3s^{-1} (1961) to 6.8 m^3s^{-1} (1990). Average flow in February's low flow conditions had declined from 9 m^3s^{-1} (1960s) to 4.59 m^3s^{-1} (1970s) to 1.29 m^3s^{-1} (1980s) and to 0.99 m^3s^{-1} (1990s).

Such problems are not restricted to Africa. In Uttaranchal's Kumaun Himalaya, many worry about declining recharge to ground waters and drought (cf. Chadha, 1989). In Kumaun, overgrazing by cattle is a problem for up to 7 km around each settlement and more than 15% of all Kumaun's watersheds are considered 'degraded' (cf. Stocking and Murnaghan, 2001). Research has shown that, while runoff from forest land may be very low, that from barren land may be three times and from cultivated land six times greater causing the water table to recede through lack of recharge (cf. Shaxson, 1999). Studies near Nainital found that 45-46% of springs had dried up or become seasonal in recent memory (Valdiya, 1985). Tiwari and Joshi's new study in the Shail catchment, Upper Kosi Basin, found that during the last 20-25 years, 18 out of 49 natural springs had completely dried up, all nine flourmills abandoned due to low water discharge, and stream heads had migrated down valley an average of 37 metres.

Despite this, they argued that there is massive under-utilisation, with less than 7% of the water resource being used. Their case study found that while the springs were capable of generating a dry season flow of 210 litre ha^{-1} day^{-1}, local people were unable to obtain the five litre day^{-1} needed for domestic use and that many fields that would benefit from irrigation received no water. Currently, irrigation waters reach only 11% of Kumaun's agricultural land. Tiwari and Joshi argue for more holistic styles of water resource development, and the creation of a Natural Resources Information System (NRIS) at micro-watershed level.

Back in Kenya, Shisanya and Kwena reviewed the ways that rural families access their domestic water supply in Kakamega District. This study of 300 respondents found that daily water requirements per capita average 120 l/day. Collection involved travel to a water point some 0.5-3.5 km from home. Most water collectors (56%) travelled less than 0.5 km but on average, each household made five trips. The average time expended in water collection was 127 minutes per day per household. Shisanya and Kwena point out that high opportunity costs, survey respondents could earn 50-60 Ksh for the same two hours spent as a casual labourer. Focus group discussions also emphasised that the empowerment of women is an important prerequisite when dealing with issues like the accessibility of water and water quality.

During the International Drinking Water Supply and Sanitation Decade (1981-1990) efforts were made by the Kenyan government and some NGOs to protect some 690 springs in Kakamega. Unfortunately, management and maintenance has since been neglected resulting in the collapse of some and contamination of others with agro-chemical residues, household garbage and carrion. Such matters motivated Items 22, 16 and 19 of the Nairobi Declaration (Janksy et al., 2003).

Special Cases: Rangelands and Wetlands in Headwater Regions

Rangelands are lands that, while unsuited for cultivation, function as sources of forage for free ranging animals. They are widespread in the Himalaya as in other headwater areas. They provide habitats for many plant species, some quite endangered, and including some of medicinal importance (Rajwar, 1993). These lands are reserves of biodiversity and also of cultural diversity, through their special pastoral communities. They attract visitors, fuelling a tourist industry that can improve local livelihoods and economic development. They also have important impacts on water quality, water supply for irrigation downstream and for sustainable power generation through hydroelectricity.

For Rajwar, range management is about *"optimising biomass productivity for the benefit of society through the manipulation of resources: biological, physical and social"*. This is explored through research on the productivity of rangelands used by six villages in the Nanda Devi Biosphere Reserve's buffer zone, Uttaranchal. Here, many people are semi-nomadic; in only two villages did the people remain inside the Biosphere Reserve buffer zone throughout the year while one village had a migrating distance of more than 40 km. Despite this, significant overgrazing tended to be a problem within 3 km of the village. Rajwar proposes that land management should be based on determination of forage production and carrying capacity and that grazing access should be restricted to minimise damage to pastures and biodiversity (cf. Joshi and Srivastava, 1991). He also commends greater use of crop residues as fodder and a search for ways of reducing conflicts between the pastoralists and wildlife. As elsewhere in these headwaters, all the villages were in decline with each year there being fewer occupied houses in the summer. The reason was the emigration of younger people, gone to seek work in the towns and lowlands. So Rajwar argues that the social conditions of the pastoralists and their problems should be considered in management plans. However, he is able to offer no immediate remedies and the real conclusion is that it seems easier to affect sustainability in the biophysical than in the social component of these headwaters.

In the Nile headwaters, wetlands are a small (1.4-5.0%) part of the total land area but very significant to their sustainable development. They provide critical water storage that helps sustain the stream flow and are vital elements in the local agroecosystem. Here, the management challenge is balancing local needs while preserving hydrological functionality.

Unfortunately, agricultural development often results in wetland drainage and their subsequent degradation. A lowered water table makes crop cultivation possible but it also accelerates runoff and erosion and reduces groundwater recharge. Eventually, desiccation causes the oxidation of humus, soil compaction and degradation. These changes affect the livelihoods of not only local cultivators but, as elsewhere, those affected by the hydrological changes downstream (cf. Krecek and Haigh, 2006).

Wood argues that wetlands become more important as land degradation reduces the capacity of surrounding interfluves to absorb rainwater but, still, little attention has been given to their environmental role. An appreciation of this led to Item 18 in the Nairobi Declaration (Box 1 and Jansky et al., 2003; Haigh, 2004). This launched a major workshop that explored the role of headwater wetlands world wide. This workshop was also dominated by scientific reports and eventually came to a social scientific conclusion—empowering local communities for effective environmental management was the best way ahead (Box 2).

Box 2: Exploring the Role of Wetlands in a Headwater Context

Results from the NATO Advanced Research Workshop 979810: *Environmental Role of Wetlands*, Marienbad, Czech Republic, December 2003 (Krecek and Haigh, 2006).

This meeting was called to solve an important practical problem for watershed managers: How to manage watersheds that contained significant wetland areas, especially those in the upper catchment and headwater areas. The problem is compounded by the fact that much current research into wetlands is difficult to apply to watershed management because it has tended to focus on the internal dynamics of the wetland – as a resource for biodiversity, economic development or hydrological phenomena – rather more than the issues that help place wetlands in their wider contexts. This problem was earlier identified in the *Nairobi Declaration for the International Year of Freshwaters 2003* (Jansky et al., 2003), which provided the mandate for the organisation of this NATO Advanced Research Workshop, which was convened in Marienbad, Czech Republic with major assistance from the International Association on Headwater Control.

Several aspects of the Nairobi Declaration were explored by this meeting (Jansky et al., 2003; Haigh, 2004). These included Item 13: *"Sustainable development should be the baseline for all environmental policy, planning, management, practice, education and law, in headwater regions"*; Item 17: *"Greater effort should be devoted to the refinement of methods for generating and sharing the appropriate and reliable information..."*; and Item 18: *"Greater attention needs to be paid to the special roles and hydrological functions of headwater wetlands and peatlands ...and also to the impacts of anthropogenic processes on watershed functions in headwater regions"*.

An assessment of the area of headwater wetlands was carried out using CORINE data for the EU15 Countries (excluding Sweden) plus Poland, Estonia, Latvia, Lithuania, Czech Republic, Slovakia, Slovenia, Hungary, Bulgaria, Romania, Bosnia and Herzegovina, Albania and Macedonia. Here, headwaters amount to 1,000,000 ha (27%), and wetlands in headwaters to 10,600 ha, about 1% of the headwater area. In the north, however, the situation is very different. Wetlands cover 14% (1.48×10^6 km^2) of Canada and greater proportions of nations like Estonia and Belarus.

Workshop discussion foci included the sustainable management of the peat lands of headwater regions, especially in northern Europe and particularly with respect to the role of hydrology, drainage and forestation, the interactions between wetlands and climatic change, the utility of wetlands in the management of environmental hazards, and the role played by stakeholders, especially the local resident communities of these regions.

Results from hydrological studies are ambiguous. The traditional view, that peat lands adsorb and store rainwater, is contested. Undisturbed peat wetlands are poor aquifers. They act mainly as surface flow systems that do little to moderate floods unless they are drained. Drainage of peat land results in drier soil covers and reduced evaporation. This, in turn, generates higher annual runoff. After drainage, peat land may moderate flood flow in dry conditions because of increased soil storage but as time passes after drainage, this effect diminishes. Eventually, peat compaction and decomposition cause the runoff to increase again. Drainage ditches also increase the speed of runoff and encourage erosion. However, the main hydrological effect of mire drainage is related more to changes in the pathways of water movement through the subsurface more than changes in water balance. Forestation does change the water balance, evapotranspiration increases and runoff is reduced. Snowmelt runoff also falls due to decreased snow melting rates and reduced snow accumulation. Overgrazing, trampling and burning encourage accelerated runoff and erosion. Coal mining is a factor in the destruction of wetlands and through subsidence, in the creation of new wetlands. Peat headwaters produce storm runoff that is acidic but drainage increases the chemical loading of most ions and reduces pH. Burning moderates pH but forestation encourages acidification. In northern Canada's permafrost area, warming by 1°C has occurred during the past 13 years where, in the same period, air temperatures have cooled or remained stable. Since the depth of permafrost in wetlands/peat lands can exceed 12 m, thawing increases the water storage capacity by as much as 60% in one case study area. This could result in reduced peak flows following snowmelt and high rainfall events and increased year-round base flows in affected basins.

Disturbed peat soils are sinks for methane while un-drained peat lands are sources. However, in general, peat lands are net sinks for C, N and many other environmental chemicals, although environmental changes can reverse this effect. Blanket peat lands are often sources for sediments, as in the UK, but most headwater wetlands tend to be net sinks for sediments as well as nutrients – as demonstrated by the Hula Valley wetland of Israel. Newly constructed headwater wetlands can be used to decrease sedimentation—as at Madan Reservoir, Bulgaria. However, small landslide-dammed lakes in mountains undergo rapid eutrophication and are complex little understood wetland habitats. The control of eutrophication processes by planktonic communities was explored in a case study at Dragichevo, Bulgaria.

Intense siltation is a major problem for environmental management around Alleghe Lake, Italy, an alpine landslide-dammed lake formed in 1771. Present problems arise from the need to remove and relocate deposited sediment in order to restore satisfactory environmental conditions and hydrological functions. The catastrophic impacts of landslide/mudflow activity and the possibilities of landslide forecasting were discussed in the context of the Log pod Mangartom mudflow of the Upper Soča Valley, Slovenia.

The debate concerning the role and function of Africa headwater wetlands echoes that in Europe. African headwater wetlands, 'dambos', play a significant role in the hydrological regime of rivers. Dambos are seasonally waterlogged grass-covered, treeless areas that remain wet in the dry season because of slow subsurface drainage from the upland areas between dambos. Outside dambos, much precipitation is lost as evapotranspiration and the land is relatively dry. Despite some scientific assertion to the contrary, dambos help conserve rainwater and sustain dry season river flows. In western Ethiopia, local communities gain benefits from such wetlands that include thatching materials, medicinal plants, spring water, seasonal grazing, craft materials and arable land. Here, demographic, economic and land use pressures are leading to increased agricultural use of wetlands and, in many cases, to their degradation. However, potentially sustainable management regimes with mixed land use have been developed, which seem to offer possibilities of maintaining many of the hydrological functions of these areas and simultaneously meeting community needs. It should be possible to construct sustainable land use regimes based on local knowledge of wetland management and local land management institutions.

Community education will be a critical aspect of promoting sustainable development in headwater regions. There is a need to ensure that the concerns and practices of water resources management were more widely and formally included in the new educational agenda being promoted in connection with the launch of the *United Nations' Decade of Education for Sustainable Development* (2005-2014). Two models of ESD were discussed. The NGO Earthwatch model promotes public education by sponsoring the engagement of community volunteers in scientific research projects. By contrast, the environmental education centre in Galtur, Austria Alps stresses the dual role the headwater as a resource and a hazard and considers how this approach can be transferred to other headwater areas. The Pian di Spagna wetland, Italian Alps, is case in point. A flood prone area until 1858 when the Adda River was canalized, this became a Ramsar reserve in 1996 because of its ecological value. However, the area is affected by critical concentrations of chromium and little attention has been paid to the rehabilitation of the flood prone areas while local vulnerability is increasing.

It is concluded that, in general, there is a need for better coordination between those planning sector-specific environmental protection measures. In addition, better environmental management understanding is required to maximise the benefits to stakeholders in upland wetlands and minimise potential negative impacts due to emissions of greenhouse gases, hydrological changes involving flooding, water chemistry, and sedimentation. There remains a need to assess the continuing role of these wetlands in current and future land use systems, especially agriculture, forestry, water resource management, tourism, and nature conservation.

The final plenary discussion commended the following action—work towards improving access to relevant data sources internationally, activities that would help keep this interest group linked and interacting via further conferences and publications, plus greater and more sustained engagement with local communities and stakeholders. It was agreed that effective environmental management for sustainability should be led by the local communities whereever possible, but this had implications for the provision of environmental education.

Beyond this, the following broad conclusions were adopted. It was agreed that there was a major underestimation of the roles played by wetlands in headwater regions in most national action plans. A new inventory was needed, perhaps constructed on the mapping methodologies described by Paracchini and Vogt (2006). In contrast to lowland wetlands, the hydrological role of headwater wetlands is neither fully understood nor properly recognised. Unfortunately, the existing system controlling wetlands (namely the conservation and protection of RAMSAR National Sites) tends to treat wetlands in isolation rather than in context and emphasises their biodiversity role more than their wider environmental functions. Better and more effective management requires better understanding of the complete role they play and the holistic integration of their attributes—biological, hydrological and socio-economic—into watershed management planning. Enhancing the husbandry of headwater wetlands within integrated watershed programmes may be achieved through the following measures: better assessment of the critical attributes of headwater wetlands that control their various functions, better community and citizen education that might help drive more effective participatory processes in watershed management, and by granting wetlands greater emphasis in planning processes.

However, Dixon and Wood go beyond such notions by recognising the sophisticated systems developed by some local communities for wetland management. However, these systems emerge from a long term process of trial and error and have proved that they work. For example, the Tulla system of the Oromo people appointed a community elder as Abba Laga or "water father" who was empowered to organise drainage and guarding, and to remove farmers who did not farm appropriately. Recently, wetland drainage schemes, cash-cropping and cattle grazing have expanded but, despite the intensification of land use, land degradation problems are associated more with places where the management schemes are recent or externally imposed and less with long-standing indigenous systems.

Although the Ramsar Convention advocates the principles of 'wise use' and 'sustainable utilisation' of wetland resources (Ramsar Convention Bureau, 2000), until recently, 'community involvement' was seen as synonymous with community co-operation in achieving externally defined conservation-orientated management goals, rather than in the incorporation of indigenous knowledge and empowerment of community control over wetland management. The challenge ahead is to help successful local wetland management systems keep pace with the changing pressures from outside. To achieve this, these institutions must be strengthened to include all stakeholders and, possibly, to define a role for external expertise in strengthening the capacity of the local communities to develop their own solutions to problems. Dixon cautions that, in future, indigenous systems may struggle to cope with the pace of accelerating environmental or socio-economic changes. However, the guiding principle must be that local communities are allowed to gain benefits, such as perennial clean spring water and craft or medicinal materials, from their wetlands so that they have vested interests in protecting these areas.

Summing Up

The discussions of IHC5 are abstracted as the *Nairobi Declaration for the International Year of Freshwaters 2003* (Janksy et al., 2003; Haigh, 2004). However, this additional sampling of reports raises several issues (Jansky et al., 2006).

Water supply is a major linking theme (*Nairobi Declaration*, Items 2-8 and 11). Several paper reports express concern about the decline of flows in rivers and springs and increasing aridity, while that of Shisanya and Kwena demonstrates the heavy cost of daily water collection. Several papers search for the impacts of global warming and secular climatic change. However, at the level of the individual catchment, the places and scales where people actually live, the main process that drives negative environmental change is the human impact that has resulted in land degradation and desiccation. Three aspects of the human impact are implicated. First, population expansion has caused deforestation, scrub clearance and soil degradation processes that reduce groundwater recharge. Second is an increasing demand for water for irrigation and other uses, which has led to excessive groundwater depletion. Together, these cause declining river flow and especially severe reductions of flow in low flow conditions. A third factor is the condition of traditional management systems. These engage large amounts of experience and expertise for the sustainable management of headwater resources. Where the systems are working, they have protected the environment from processes that have caused degradation elsewhere. Unfortunately, some of the systems examined are at, or beyond, their limits. The rangeland pastoralist communities of Uttaranchal are suffering rapid decline due to emigration. Those in Ethiopia are in danger of being overwhelmed by the pace and scale of social and economic change.

All of these reports underline the vital need for researchers to work with the community of land users to promote better land husbandry and to help those communities help themselves to adapt to a changing world. Shaxson et al. (1997, p. 1) talk about *"collaborative efforts which are interdisciplinary, supportive of farmers' aspirations, and based on joint learning"* (cf. *Nairobi Declaration,* Items 16, 22, and 23). This conference, part of the evolution of the Headwater Control philosophy, also marked as the moment when that movement publicly changed its aspiration, from controlling headwater habitats to fostering headwater *swaraj*, headwater self-control.

Wetlands were a special concern of this conference and the subject of a major follow-up workshop (*Nairobi Declaration* Item 18; Box 1). The upshot of this was agreement that there the roles of wetlands in headwater regions was underplayed in most national action plans. An inventory was needed (cf. Paracchini and Vogt, 2006). Meanwhile, the hydrological role of headwater wetlands is not fully understood while, the existing system controlling wetlands (RAMSAR National Sites), which tends to treat wetlands in isolation rather than in context, is unhelpful. Better and more effective management requires more holistic integration of the many wetland functions— biological, hydrological and especially socio-economic—into watershed management planning. Enhancing the husbandry of headwater wetlands also requires better education to help drive more effective community-empowered watershed management.

Today, as Kofi Annan's Dhaka speech argued, *"Our biggest challenge... is to take an idea that sounds abstract – sustainable development – and turn it into reality for all the world's people"* (Annan, 2001, p. 2). From the outset, the Headwater Control Movement has sought to unite the perspectives of the researcher, the field practitioner and the policy maker. Hellin (2006) suggests that, today, the pendulum may have swung too far from the biophysical aspects of sustainable development and towards the social and economic. However, the papers from this Nairobi Conference suggest that there is still a great deal more to be done. The problem may be less the effort devoted to the scientific study of headwaters, than the mind sets of those who undertake the research. In brief, too little of the work done is directly relevant to the problems of sustainable development. Too many researchers are content to produce results of the vague academic significance. Too few bother to ground truth their work by finding out what the people actually embedded in the development—hopefully sustainable development—of headwater regions actually need.

India long nurtured a notion that the best people are those who retreat from the world of action. Like the 'elite' Western academic, those who aspire to the path of knowledge think high thoughts but remain aloof from practical labour. However, from Swami Vivekananda, Mahatma Gandhi and Acharya Vinoba Bhave, India's intellectual leaders commend the life of engagement and service undertaken from a sense of duty, without attachment to the fruits of that action, the path of karma (Vivekananda, 1907; Bhave, 1986). It is a model the world could adopt.

Meanwhile, our problems are magnified by a dominant globalised culture based on highly suspect ideas. For example, Thomas Berry (1999, p. 102) notes: *"Our legal system fosters a sense of human rights, with other than human beings having no inherent rights. Our economic system is based on the exploitation of the Earth in all its geobiological systems. Commercial rights to profit prevail over urgent need of natural systems for survival"*. Genuine sustainability will require that our society disengages from such thinking. At present, nothing takes priority over our immediate social needs or human values. Berry (1999, p. 56) argues that our route towards becoming a viable species is *"to shift from a human-centred to an Earth-centred norm of reality and value"*. This is not a negation of concern for humans but enlightened self-interest that seeks to preserve the whole human habitat. As Kofi Annan (2004, p. 1) points out: *"Our highest priority should be to guarantee the health and effective functioning of the Earth's life support systems"*. Rationally, this seems self-evident, yet the reality revealed in the media seems different.

Kofi Annan (2004, p. 1) also notes that: *"Every individual... has an obligation and an interest in changing outlooks through education and by example, thereby helping to end thoughtless or deliberate waste and destruction"*. The Nairobi Declaration (Item 20) anticipates the launch of the United Nations Decade of Education for Sustainable Development 2005-2014 (UN, 2004) by recognising that sustainable development is not merely an issue for professionals in environmental management and research but for every individual in the community (Haigh, 2006). Making sustainable development a reality, this task, more than any science, technology or even social science, is the real front-line for work in headwater areas. Here, the results of IHC5 suggest that there are four arbiters for success: a clear and agreed message, clearly targeted action, a willingness to work with others (not least the UN agencies), and personal commitment as the main motivation. Ervin Laszlo comments: *"The world needs changing. It is not sustainable, and sooner or later it will change. Predicting which way it will change is not the challenge before us, the future is not to be foretold, it is to be created. The challenge is to create a positive future. And that is up to us – to you and me"* (Laszlo, 2002, p. 1).

References

Agrawal, A., 1995. Dismantling the divide between indigenous and scientific knowledge. *Development and Change,* **26:** 413-439.

Annan, K., 2004. Secretary General says failure to stop biodiversity loss too awful to contemplate. *United Nations Press Release* SG/SM/9289, ENV/DEV/780,OBV/420 (03/05/2004) (http://www.un.org/News/Press/docs/2004/sgsm9289.doc.htm - Accessed June 2004).

Annan, K., 2001. Secretary General calls for break in political stalemate over environmental issues. *United Nations Press Release*: SC/SM/7739 (15/03/01). (http://www.un.org/News/Press/docs/2001/sgsm7739.doc.htm - Accessed April 2004).

Berry, T.M., 1999. The Great Work. Bell Tower, New York.

Bhave, V., 1986. The Intimate and the Ultimate. Green Books, Totnes, UK.

Bridges, E.M., Hannam, I.D., Oldeman, L.R., Penning de Vries, F.W.T., Scherr, S. and Sombatpanit, S. (Eds), 2001. Response to Land Degradation. Science Publishers Inc., Enfield, NH.

Chadha, S.K. (Ed.), 1989. Environmental Holocaust in the Himalaya. Ashish Publishing House, New Delhi.

Chambers, N., Simmons, C. and Wackernagel, M., 2001. Sharing Nature's Interest: Ecological Footprinting as an Indicator of Sustainability. Earthscan, London (See also: ecological footprinting questions: http://www.bestfootforward.com/footfaq.html - Accessed November 2004).

Haigh, M.J., 2006. Greening the curriculum: Appraising an international movement. *Journal of Geography in Higher Education,* **29(1):** 33-50.

Haigh, M.J., 2004. Sustainable management of headwater resources: The Nairobi 'Headwater' Declaration (2002) and beyond. *Asian Journal of Water, Environment and Pollution,* **1(1-2):** 17-28. (http://www.capital-publishing.com/journal/; http://www.iospress.nl/html/09729860.html).

Haigh, M.J., Jansky, L. and Hellin, J., 2004. Headwater deforestation: a challenge for environmental management. *Global Environmental Change: Human and Policy Dimensions,* **14** (*S1: Special issue - Focus on the International Year of Freshwater*): 51-61. (www.elsevier.com/locate/gloencha).

Haigh, M.J., Krecek, J., Rajwar, G.S. and Kilmartin, M.P. (Eds), 1998. Headwaters: Water Resources and Soil Conservation. A.A. Balkema, Rotterdam.

Hellin, J., 2006. From Soil Conservation to Better Land Husbandry: Towards a Holistic Approach to Land Management. *In:* Land Reconstruction and Management 4. Science Publishers, Enfield, NH.

IPCC, 1996. Climate Change 1995: Impacts, Adaptations and Mitigation of Climate Change: Scientific-Technical Analyses. Contribution of Working Group II to the 2nd Assessment Report of the Intergovernmental Panel on Climate Change.

Jansky, L., Haigh, M.J. and Prasad, H. (Eds), 2006. Sustainable Management of Headwater Resources – Focus on Africa and India. United Nations University Press, Tokyo.

Jansky, L., Krecek, J. and Haigh, M.J., 2003. Sustainable Management of Headwater Resources (Nairobi "Headwater" Declaration for the International Year of Freshwaters, 2003). Tokyo, United Nations University. (http://update.unu.edu/downloads/declaration.doc - Accessed January 2004).

Joshi, S.P. and Srivastava, M.M., 1991. The status of grazing lands of alpine regions in the Garhwal Himalayas, pp. 1-11. *In:* Rajwar, G.S. (Ed.) Recent Researches in Ecology, Environment and Pollution 6. Scholarly Publications, Houston/Today and Tomorrow's Printers & Publishers, New Delhi.

Krecek, J. and Haigh, M.J. (Eds), 2006. Environmental Role of Headwater Wetlands. NATO/Kluwer, Dordrecht.

Krecek, J., Grip, H., Haigh, M.J. and Hocevar, A. (Eds). 1989. Headwater Control (2 vols). CSVTS, IUFRO & WASWC, Plsen, Czechoslovakia.

Krecek, J. and Haigh, M.J. (Eds), 1992. Environmental Regeneration in Headwaters: Proceedings of the Second International Conference on Headwater Control. Enco, Prague.

Laszlo, E., 2002. You Can Change the World: Action Handbook for the 21st Century-A Report from the Club of Budapest. Positive News Publishing, Clun, UK.

Paracchini, M.L. and Vogt, J.V., 2006. Mapping wetlands in European headwater areas, pp. xx-yy. *In:* Krecek, J. et al. (Eds) Environmental Role of Headwater Wetlands. NATO/Kluwer, Dordrecht.

Rajwar, G.S. (Ed.), 1993. Garhwal Himalaya: Ecology and Environment. Ashish Publishing House, New Delhi.

Ramsar Convention Bureau, 2000. Wise Use of Wetlands. Ramsar Handbook 1. Gland, Switzerland.

SAPA-AFP, 2005. IOL Environment: Count Down to Environmental Catastrophe. Independent Online (24.01.2005) (http://www.iol.co.za/index.php?set_id=14&click_id=143&art_id=qw1106541006165B216 - Accessed January, 2005; see also: http://www.newscientist.com/channel/earth/climate-change).

Schneider, S.H., 1996. Laboratory Earth: The Planetary Gamble We Can't Afford to Lose. Science Masters Series. Weidenfield and Nicolson, London.

Shaxson, T.F., 1999. New Concepts and Approaches to Land Management in the Tropics with Emphasis on Steeplands. *FAO Soils Bulletin,* **75**. Rome.

Shaxson, T.F., Tiffen, M., Wood, A. and Turton, C., 1997. Better Land Husbandry: Re-Thinking Approaches to Land Improvement and the Conservation of Water and Soil. *In:* Natural Resource Perspectives 19. Overseas Development Institute, London (http://www.odi.org.uk/nrp/19.html - Accessed June 2004).

Singh, R.B. and Haigh, M.J. (Eds), 1995. Environmental Reconstruction of Highland and Headwater Regions. A.A. Balkema, Rotterdam.

Stocking, M.A. and Murnaghan, N., 2001. Handbook for the Field Assessment of Land Degradation. Earthscan, London.

UN, 2003. Resolution 57/254. United Nations Decade of Education for Sustainable Development. Resolution adopted by the General Assembly. United Nations General Assembly, 57th Session, 21 February 2003, Agenda item 87 (a) 02 55612; Document: A/RES/57/254. United Nations, New York. (http://www.unesco.ch/pdf/UNDekade.pdf - accessed August 2003).

Valdiya, K.S., 1985. Accelerated erosion and landslide prone areas in the Central Himalayan Region. *In:* Singh, J.S. (Ed.) Environmental Regeneration in Himalaya. Central Himalayan Environment Association, Nainital.

Vivekananda, 1907. Karma yoga. *In:* The Complete Works of Swami Vivekananda 1 (Ed. 1989). Advaita Ashrama (Calcutta), Kolkata.

2

Social Science Contributions to Multiple Objective Decision Making within Watersheds

Ted L. Napier
Department of Human and Community Resource Development
The School of Natural Resources
Ohio State University, 2120 Fyffe Road
Columbus, Ohio 43210, USA

Introduction

Practically all natural resources development programmes have been implemented at the watershed level in recent years because it has been repeatedly observed that watersheds are integrated socio-environmental units whose component parts are interdependent. The well-being of the components is dependent on the viability of the whole. To achieve improvement in environmental quality in specific components of a watershed requires a holistic planning and programme implementation approach.

Attempts to address natural resources development without consideration of a multitude of other development issues within watersheds will often result in failure. It must be recognized that effective resolution of natural resource problems within one component of a watershed is partially dependent on effectively addressing multiple issues within other components. Integrated planning and intervention programming at the watershed level necessitates simultaneous consideration of natural resources and community development initiatives throughout the watershed. Creation of effective intervention strategies at the watershed level poses more complex planning and implementation problems than programmes focused on a single issue within a specific community or sub-watershed.

There are a host of natural resources and community development issues that need to be addressed when watersheds are large in size and encompass multiple communities and sub-watersheds. Problems associated with watershed planning are further compounded when topography varies within the watershed and when the socio-economic and cultural compositions of resident populations differ significantly. Each community or ethnic group will mostly be primarily concerned about community-based natural resources and community development issues and will be less willing to address collective watershed problems. This is especially true when community-based economic and human resources are required to be allocated to portions of the watershed that lie outside of the local community. It is very difficult to motivate watershed groups to contribute local human and economic resources to development efforts in other areas of the watershed even when local groups will benefit directly from the investments made elsewhere. A classic example of this problem is the necessity for downstream residents to invest limited development resources in headwater regions of watersheds to accomplish natural resources and community development goals. Inevitably a significant number of downstream residents will strongly resist requests for contributions of limited resources to implement upstream natural resources and community development efforts. Watershed residents often are in direct competition for existing economic and human resources and the conflict can impede achievement of collective development goals.

Another problem associated with holistic watershed planning and programme implementation is the establishment of priorities among competing natural resources and community development options. Since economic and human resources are usually very limited for addressing collective watershed problems, some means must be devised to assess the priority attributed to each action option. It is often difficult for large and diverse populations to agree on what natural resources and community development option(s) will be emphasized within watersheds. Some watershed residents may perceive a need for agro-forestry projects to protect riparian corridors, while others may place priority on the creation of physical structures to be placed in streams to prevent stream-bank erosion and to retard stream velocities. Agriculturalists within the watershed may favour allocating limited human and economic resources to the construction of terraces to control erosion and to enhance agricultural production. Other watershed dwellers may favour planning efforts that will contribute to industrial/commercial development rather than protection of natural resources. Still others may wish to invest development resources in residential housing and improved streets. Unless there is agreement among watershed residents on the goals to be accomplished by intervention programmes, the opportunity costs associated with allocating limited resources to development efforts that are less critical to the well-being of watershed residents could be very high.

One means of dealing with complex planning problems within watersheds is multiple objective decision making that incorporates contributions from many disciplines (El-Swaify and Yakowitz, 1997). Contributions from physical and social scientists are integrated within the planning process to establish natural resources and community development objectives and to identify means of accomplishing the goals. The primary role of social scientists within the multiple objective decision making framework is to provide research findings that will advance the planning process. Some of the most significant contributions social scientists can make to multiple objective decision making within watersheds are as follows: (1) the examination of the social acceptability of alternative technological and sociological solutions to identified problems, (2) the assessment of public preferences for alternative development options, (3) the examination of alternative organizational structures required to advance development options, and (4) the evaluation of public policies formulated to manage watershed development. The purpose of this paper is to discuss contributions made by social scientists to each of these research traditions.

Contributions of Social Scientists to Technology Transfer of Conservation Innovations

One of the most significant contributions of social scientists to multiple objective decision-making at the watershed level is research focused on technology transfer of natural resources conservation technologies and techniques. While physical scientists can undoubtedly develop technologies and/or techniques that will resolve nearly any environmental problem within watersheds, the social acceptability of many potentially feasible solutions is often problematic. One of the primary contributions that social scientists can make to watershed planning is to identify why some residents are willing to adopt specific natural resources conservation innovations while rejecting others. Such information can be employed to identify technological and sociological innovations that will be defined as being acceptable by watershed residents. Such information can aid planners in the selection of conservation and community development options that will generate the greatest support among directly affected populations.

Social scientists have been very active in assessing adoption of soil and water conservation production systems among watershed land managers during the past 25 years (Batte and Bacon, 1995; Bridges and Napier, 2003; Camboni and Napier, 1993; Halcrow et al., 1982; Lasley et al., 1990; Napier et al., 1983, 1994, 2004; Lovejoy and Napier, 1986, 2000; Mueller et al., 1985; Napier and Bridges, 2002, 2003; Napier and Tucker, 2001; Putman and Alt, 1987; Robinson and Napier, 2002; Swanson and Clearfield, 1994). This body of research has clearly demonstrated that many social, economic, farm structure, and public policy variables, commonly argued to be significantly correlated with adoption of conservation production systems in the US, are not very useful for motivating land owner-operators to adopt and to use soil and water conservation production systems at the farm level. Factors such as access to conservation information and training, government-sponsored technical assistance, access to government cost-sharing, characteristics of the farm enterprise, farm specialization, psychosocial orientations toward the environment and protection of natural resources, and a host of other socio-economic factors have been shown to have very limited predictive utility relative to measures of adoption and use of soil and water conservation production systems at the farm level. Similar findings have been reported in other societies (Napier et al., 1994, 2000).

The implication of these findings for natural resources conservation efforts at the watershed level in the US and abroad is that standardized approaches designed to address degradation of soil and water resources will probably not be successful. Study findings strongly indicate that no single policy and/or implementation approach will be widely applicable to large and diverse populations. This is especially true for natural resources conservation programmes that emphasize the provision of conservation information, conservation training and partial subsidies (Napier et al., 2004). While such a conservation approach may motivate a small segment of client populations to adopt conservation production systems at the farm level, it will not resolve most natural resources conservation problems at the watershed level. Unfortunately, education and information programmes have been employed extensively throughout the world and have been acclaimed to be the best means of encouraging resource managers to adopt and use conservation production systems at the farm level. Research evidence generated by social scientists has demonstrated that such a strategy will not be successful in the long term and perhaps not in the short term. While land managers must possess skills to use conservation production systems, they also must have access to economic resources and conservation technologies to address most natural resources development problems.

The only factor that has been shown to consistently facilitate the adoption and continued use of conservation production systems is economic subsidies (Napier et al., 2004; Napier, 2005a). It has also been observed that economic subsidies must be sufficiently high to make it profitable for land owner-operators to adopt and use conservation production systems. Large economic subsidies will motivate farmers to adopt and use conservation technologies and techniques because land managers are not required to assume the financial risks associated with adopting innovations that may result in financial loss. The magnitude of the monetary compensation required to motivate land owner-operators to adopt conservation systems will vary from farm to farm and from one sub-watershed to another within a watershed. While many of the studies focused on adoption of conservation production systems have been conducted in high-scale societies, it is almost certain that the same factors will operate in the same manner in lesser-scale societies (Napier et al., 1994).

Research findings focused on the role of economic resources as motivators of adoption behaviours strongly suggest that soil and water conservation programmes initiated within watersheds must have continuity of funding and that the magnitude of funding must be sufficiently high to motivate land managers to adopt. Soil and water conservation programmes implemented within watersheds without adequate economic resources to provide large subsidies to participating land managers are almost certain to be doomed to failure.

Research findings derived from studies focused on the adoption of soil and water conservation production systems at the farm level strongly indicate that social scientists should determine the magnitude of economic subsidies that will be required to motivate land managers to adopt conservation production systems. Social scientists should also be involved in the determination of the need for technical assistance and training that is required to effectively implement soil and water conservation programmes at the farm level. Social science findings from such studies will provide insight to the feasibility of initiating soil and water conservation efforts within specific watersheds.

Assessment of Public Preferences

Another significant social science contribution to multiple objective decision-making at the watershed level is the assessment of public preferences for alternative natural resources development options. One of the most significant barriers to public participation in watershed planning and programme implementation is the difficulty associated with involving large and diverse populations in a meaningful manner. Some watershed residents will be willing to participate in the selection of natural resources and community development goals and contribute time and economic resources to achieve policy goals and programme objectives. Unfortunately, a large percentage of watershed residents are either ambivalent about participating in development efforts or do not wish to do so. An important task of social scientists is to conduct research to assess the development preferences of watershed residents who do not wish to become directly involved in the decision making process so that preferences of nonparticipants are reflected in the establishment of goals and implementation strategies.

One means of involving a broad cross-section of watershed residents in the decision making process is via the use of survey research (Napier, 2003). Social surveys can produce findings that are representative of diverse watershed populations, if the samples are scientifically drawn using random sampling techniques and if the number of people included in the study sample is large enough to produce a study group that is representative of the watershed. Findings generated using nonrepresentative samples can be counter-productive relative to addressing natural resources and community development at the watershed level because priorities may be established that do not reflect the perceived need of watershed residents. The size of the sample must be quite large, if the population being sampled is diverse in terms of socio-economic and cultural factors. People from all socio-economic classes and ethnic groups must be included in the sampling to produce valid and reliable results for planning purposes and for effective programme implementation.

Some social scientists have been trained in the scientific methodologies of survey research and findings produced by trained professionals can contribute significantly to watershed planning and programme implementation. Survey research can identify development priorities among diverse populations and can determine the level of political support among watershed residents for alternative natural resources development options. Research can also determine the preferences of specific sub-groups within the watershed whose special needs must be incorporated into the planning process. This is especially true for indigent and ethnic groups who constitute a minority within the watershed.

One of the most useful approaches for establishing priorities for natural resources and community development is to provide selected respondents with realistic alternative development options and to ask them to rank order the options in terms of priorities. Several natural resources and community development problems that could be effectively addressed by watershed groups can be provided and respondents can be asked to indicate the relative importance of each development option. Respondents can also be asked to indicate the type of economic and human resources that should be allocated to each action option (Napier, 2003). Such information can provide watershed planners with information for choosing among development options that are socially acceptable to client populations. Client populations have a much higher probability of supporting development efforts that are consistent with their preferences.

Another contribution of social science to natural resources and community development planning and programme implementation is the assessment of support or opposition to proposed development efforts. The ultimate goal of multiple objective decision-making is to formulate policies and programmes that will improve the well-being of resident populations within the watersheds. All well-intentioned planning groups can make errors and produce development policies and create programmes that are not socially acceptable to directly affected populations. Social science research can assess psycho-social orientations of resident populations toward proposed development efforts and the results can provide confirmation of the planning process or identify significant opposition to what is being proposed (Sullivan and Napier, 2005).

Identification of significant opposition to proposed natural resources and community development policies and programmes early in the planning and programme implementation stages makes it possible for planners to consider more acceptable development options. Evaluation of development proposals can also result in the rejection of public development policy and programmes by local populations (Sullivan and Napier, 2005). Political action to block implementation of development policies and programmes found to be unacceptable to local people can prevent ill conceived natural resources and community development efforts from being implemented.

Examination of Organizational Structures to Achieve Development Goals and Objectives

Effective natural resources and community development efforts within watersheds require cooperation among many diverse groups and one of the most significant means of accomplishing this requisite function is via formal and informal organizational structures (Napier, 1998a; Koontz and Johnson, 2004). Some social scientists have developed skills in the assessment of organizational needs and performance. Such skills can facilitate the establishment of effective local, regional and national organizations to participate in the planning and implementation of watershed natural resources and community development policies and programmes. Monitoring of organizational functioning over time can contribute to modifications being made in organizational structures and functions to better achieve development goals and objectives.

Natural resources and community development efforts cannot be expected to achieve desired outcomes without a high level of cooperation among various watershed interest groups. Successful development efforts at the watershed level require economic resources (Born and Genskow, 2000), human resources (Fleishman, 2004), linkages to external organizations (Koontz et al., 2004; Moseley and KenCarin, 2000), technical expertise (Napier and Johnson, 1998), inter-organization cooperation (Koontz et al., 2004), and a host of other needed inputs. While availability of these resources will not guarantee successful achievement of development goals and objectives (Napier and Johnson, 1998), it is certain that inability to access these types of resources will contribute to failure. To satisfy these requisite needs for successful development outcomes, watershed planners and programme implementors must facilitate the organization of watershed groups to address each of these functions. The effectiveness of this process will be a major contributor to success or failure of development efforts (Sullivan and Napier, 1998, 2005, 2005b; Napier and Johnson, 1998).

Monitoring of local organizational structures and functions over time is also a very important contribution of social scientists to effective watershed management. Such research activities make it possible for planners and programme implementors to modify existing organizations to meet new challenges and to replace ineffective group members. Human beings in all organizations frequently become so immersed in the functioning of the group that they place more emphasis on the survival of the organization than they do on the achievement of natural resources and community development goals that brought the organization into being. Social scientists can monitor organizations and prevent goal displacement from impeding the accomplishment of natural resources and community development goals and objectives.

Assessment of Public Conservation Policies

Successful natural resources and conservation development initiatives within large watersheds almost always require public policies to facilitate the coordination of action programmes and to authorize sufficient funding to accomplish programme goals. One of the major contributions of social scientists to multiple objective decision-making is to determine what types of public policies are socially and politically acceptable to resident populations. Some public policies are perceived positively by directly affected groups while others are viewed negatively.

Research evidence strongly indicates that watershed groups favour voluntary and incentive-driven natural resources and community development policies and programmes (Napier, 1998b, 1998c, 2004, 2005a; Sullivan and Napier, 2005). In the context of natural resources development efforts, such policies and programmes provide incentives to land managers to participate in conservation efforts at the farm level by providing technical assistance and economic subsidies. In many instances, these types of policy approaches are not successful in permanently resolving environmental degradation because land managers often perceive such incentives as payments for actions taken at a specific place and time. When the incentives are withdrawn, land resources are most often returned to production using the same environmentally degrading production systems that were employed before the land manager participated in the conservation efforts. In many instances, conservation benefits achieved by incentive-driven conservation programmes are lost once participation in the programme ceases (Napier, 2005).

This body of research strongly indicates that incentive-driven approaches to natural resources development will probably not be very successful because permanent resolution of environmental problems is seldom achieved. Subsidies become quasi-transfer payments to the agricultural sector and are perceived as rents. Politically, such actions are strongly supported by land owner-operators because they are the recipients of the governmental subsidies and are not required to retain natural resources conservation production systems once programme participation ceases. Such conservation policy approaches basically legitimize the continued subsidization of land owners.

A useful contribution of social scientists to multiple objective decision-making at the watershed level is to provide planners and policy makers with evaluations concerning the effectiveness of alternative natural resources and community development policies (Napier et al., 1994, 2000, 2004; Napier and Johnson, 1998). Extensive research has revealed that many natural resources development policies have not been successful in achieving stated objectives. For example, examination of the benefits and costs of the Conservation Title of 1985 indicated that economic costs were high and that expected benefits were not distributed in a manner deemed appropriate. Evaluation research findings focused on the effects of the Conservation Title of 1985 were used to modify subsequent Conservation Titles contained within the 1990, 1996 and the 2002 Farm Bills. The modifications made in Conservation Title policies and programmes produced outcomes that were more consistent with environmental goals established by society. These findings suggest that social scientists should be actively involved in watershed management after public policies have been created and natural resources programmes have been implemented because policies and programmes may need to be modified to achieve goals and objectives.

Assessment studies focused on voluntary and incentive-driven conservation policies and programmes strongly suggest that other intervention approaches will have to be considered in the future if environmental degradation is to be permanently addressed. Assessments of policy approaches in several countries in Europe suggest that command and control strategies may have merit (Napier et al., 1994, 2000). Social scientists often possess the research skills to assess the effectiveness of command and control approaches for accomplishing natural resources and community development goals and objectives. These findings should be useful for formulating more relevant public natural resources and community development policies and programmes in the future.

Summary and Conclusion

Multiple objective decision-making at the watershed level poses significant challenges and unique opportunities to address critical socio-environmental problems for resident populations. Both physical and social scientists have significant roles to play in the decision-making process. However, success of future development efforts will hinge on the willingness of social and physical scientists to become intimately integrated into research/development/planning teams that are highly interdisciplinary in nature. The era of independent professional entrepreneurial activities for specialized disciplines in natural resources and community development is a thing of the past. Social scientists can aid physical scientists in the research and development of needed technologies and techniques to address environmental issues by identifying the types of innovations that will be acceptable to client groups. Physical scientists can aid social scientists in their research by identifying the feasibility of creating innovations to address specific socio-environmental problems. Such cooperation will make the research output for both the physical and social scientists more relevant to the needs of client groups and will reduce the incidence of irrelevant research. Physical scientists will not be required to expend human and economic resources in the research and development of technological innovations that will not be accepted by client groups. Social scientists will not be required to assess innovations that are not technologically feasible options.

Social scientists in the future will be required to cooperate more closely than in the past with other social science disciplines to accomplish watershed management goals and objectives. Political scientists, sociologists, economists, educators, planners, and other social science disciplines must move beyond discipline boundaries and seize the opportunity to participate in integrated research and planning opportunities. Each discipline can make contributions to the decision-making process by facilitating multiple objective decision-making at the watershed level.

No single physical or social science discipline possesses the necessary knowledge base or research skills to address the multitude of complex socio-economic and physical science problems posed by integrated watershed planning and development. However, interdisciplinary cooperation offers local, regional, national and international natural resources and community development professionals the opportunity to collectively address socio-environmental problems that seriously threaten the socio-environmental well-being of human beings throughout the world.

Acknowledgement

Funding for the development of this manuscript was provided by the Ohio Agricultural Research and Development Center of the Ohio State University. The positions advanced in the article are solely those of the author and do not represent official positions of the Ohio State University.

References

Batte, M.T. and Bacon, K.L., 1995. Economic evaluation of three production regimes at the Ohio MSEA project. *In:* Clean Water-Clean Environment, 21st Century. American Society of Agricultural Engineers, St. Joseph, Michigan.

Bridges, T.R. and Napier, T.L., 2003. Factors that influence farm women to advocate the adoption of environmentally benign agricultural production practices. *International Journal of Agricultural Resources, Governance and Ecology,* **2(2):** 201-219.

Camboni, S.M. and Napier, T.L., 1993. Factors affecting use of conservation practices in east central Ohio. *Agriculture, Ecosystems and Environment,* **45:** 79-94.

El-Swaify, S.A. and Yakowitz, D.S. (Eds), 1997. Multiple Objective Decision Making for Land, Water, and Environmental Management. Lewis Publishing Company, Boca Raton, Florida.

Fleishman, R., 2004. Watershed Groups in Ohio: The Effects of Organizational Characteristics on political behavior, Accomplishments, and Perceived Effectiveness. Master Thesis. The Ohio State University, Columbus, Ohio.

Halcrow, H.G., Heady, E.O. and Cotner, M.L., 1982. Soil Conservation Policies, Institutions, and Incentives. Soil and Water Conservation Society Press, Ankeny, Iowa.

Koontz, T.M. and Johnson, E.M., 2004. One size does not fit all: Matching breath of stakeholder participation to watershed group accomplishments. *Policy Sciences,* **37:** 185-204.

Koontz, T.M., Steelman, T.A., Carmin, J., Korfmacher, K.S., Moseley, C. and Thomas, C.W., 2002. Collaborative Environmental Management: What Roles for Government? Resources for the Future, Washington, D.C.

Lasley, P., Duffy, M., Kettner, K. and Chase, C., 1990. Factors affecting farmer's use of practices to reduce commercial fertilizers and pesticides. *Journal of Soil and Water Conservation,* **45(1):** 132-136.

Lovejoy, S.B. and Napier, T.L., 1986. Conserving Soil: Insights from Socioeconomic Research. Soil and Water Conservation Society Press, Ankeny, Iowa.

Moseley, C. and KenCairn, B., 2000. Institutional problem solving or social change: The Applegate and Grand Canyon Forest Partnerships. Paper presented at the Steps Toward Stewardship Conservation Conference, April 2000. Northern Arizona University, Flagstaff, Arizona.

Mueller, D.H., Klemme, R.M. and Daniel, T.C., 1985. Short- and long-term cost comparisons of conventional tillage systems in corn production. *Journal of Soil and Water Conservation,* **40(5):** 466-470.

Napier, T.L., 1998a. Conservation coalitions cannot overcome poor conservation programming. *Journal of Soil and Water Conservation,* **53(4):** 300-303.

Napier, T.L., 1998b. Soil and water conservation policy approaches in North America, Europe, and Australia. *Water Policy,* **1(1):** 551-565.

Napier, T.L., 1998c. North American and Western Europe conservation policies: contrasts in strategies. *Conservation Voices,* **1(5):** 34-36.

Napier, T.L., 2000. Use of soil and water protection practices among farmers in the North Central Region of the United States. *Journal of the American Water Resources Association,* **36(4):** 723-735.

Napier, T.L., 2003. Muskingum River Watershed Natural Resources Development Needs Assessment Study Report, pp. 39.

Napier, T.L., 2004. Public conservation policy and agricultural sustainability. *In:* Svatos, M. and Bohackova, I. (Eds) Sustainable Development of an Agrarian Sector: Challenges and Risks. Czech Agricultural University Press, Prague, Czech Republic.

Napier, T.L., 2005a. Soil and Water Conservation Policies in the United States: Evolutionary Trends and Future Options. Accepted for publication in an edited text on *Agrodiversity*. M.A. Zoebisch (ed.). (forthcoming).

Napier, T.L., 2005b. Environmental assessment of conservation initiatives within the Darby Creek watershed in Ohio. Accepted for publication in *Monitoring and Evaluation of Soil Conservation and Watershed Development Projects*, J. de Graff, P. Christian, S. Sombatpanit and J. Cameron (eds.). IBSRAM. (forthcoming).

Napier, T.L. and Bridges, T.R., 2002. Adoption of conservation production systems in two Ohio watersheds: A comparative study. *Journal of Soil and Water Conservation,* **57(4):** 229-235.

Napier, T.L. and Bridges, T.R., 2003. Adoption of conservation production systems within the upper region of the Scioto River watershed in Ohio. *Food, Agriculture, and Environment,* **1(2):** 287-294.

Napier, T.L., Camboni, S.M. and El-Swaify, S.A., 1994. Adopting Conservation on the Farm: An International Perspective on the Socioeconomics of Soil and Water Conservation. Soil and Water Conservation Society Press, Ankeny, Iowa.

Napier, T.L., Henry, C. and Yang, X., 2004. Impacts of conservation policies and programs on farm level adoption behaviors in the United States. *Trends in Soil Science,* **3:** 47-58.

Napier, T.L., Napier, S.M. and Tvrdon, J., 2000. Soil and Water Conservation Policies and Programs: Successes and Failures. CRC Press, Boca Raton, Florida.

Napier, T.L. and Johnson, E.J., 1998. Impacts of voluntary conservation initiatives in the Darby Creek watershed of Ohio. *Journal of Soil and Water Conservation,* **53(1):** 78-84.

Napier, T.L., Scott, D., Easter, K.W. and Supalla, R. (Eds), 1983. Water Resources Research: Problems and Potentials for Agriculture and Rural Communities. Soil and Water Conservation Society Press, Ankeny, Iowa.

Napier, T.L., Thraen, C.S., Gore, A. and Goe, W.R., 1984. Factors affecting adoption of conventional and conservation tillage practices in Ohio. *Journal of Soil and Water Conservation,* **39(3):** 205-209.

Napier, T.L. and Tucker, M., 2001. Factors affecting adoption of nutrient application rates within three Midwestern watersheds. *Journal of Soil and Water Conservation,* **56(3):** 220-228.

Putman, J. and Alt, K., 1987. Erosion control: How does it change farm income? *Journal of Soil and Water Conservation,* **42(4):** 265-267.

Robinson, J.R. and Napier, T.L., 2002. Adoption of nutrient management techniques to reduce hypoxia in the Gulf of Mexico. *Agricultural Systems,* **72:** 197-213.

Swanson, L.E. and Clearfield, F.B., 1994. Agricultural Policy and the Environment: Iron Fist or Open Hand. Soil and Water Conservation Society Press, Ankeny, Iowa.

3

Managing Headwater Regions in Australia: Assessing Socio-economic and Resource Sustainability

John Cary
Institute for Sustainability and Innovation, Victoria University
PO Box 14428, Melbourne City MC 8001 Australia

Introduction

For the people of the desert were as little static as the people of the hills. The economic life of the desert was based on the supply of camels which were best bred on the rigorous upland pastures with their strong nutritive thorns. By this industry the Bedouins lived; and it in turn moulded their life, apportioned the tribal areas, and kept the clans revolving through their rote of spring, summer and winter pasturages, as the herds cropped the scanty growths of each in turn. The camel markets in Syria, Mesopotamia, and Egypt determined the population which the deserts could support, and regulated strictly their standard of living. (Lawrence 1986, p. 26)

The ecological, economic and social fragility of the Arabian desert and, in turn, its dependent uplands communities, described by T.E. Lawrence, soon becomes more complex when overlain with human activity and external economic forces, even for an uncomplicated nomadic monoculture of camel-rearing. Modern agricultural and land-use systems often reflect a many-fold increase in complexity.

Agricultural production systems, particularly those in sensitive headwater regions, will always be subject to disruptions and perturbations. Such disruptions are a reflection of ecological and social evolution which, in turn, provides the impetus for adaptation (Wilkinson and Cary, 2002). In an adaptive system approach (Holling, 1978; Lee, 1993) any equilibrium state is likely to be transitory, existing only until the next perturbation.

A sustainable ecological or social system is one that can adapt to major disturbing forces, rather than one that can withstand them. A fundamental idea in biological evolution is the difference between processes of adaptation and adaptability (Boulding, 1978). In biological evolution, species display adaptation to particular environment niches and display varying levels of adaptability to changing environments. 'Adaptation to a particular niche, . . . while it leads to short-run survival, is never adequate for survival in the long run.

Adaptability is the capacity to expand niches or to find new niches (Boulding, 1978). Because sustainable states evolve (and often subsequently fail), it is often much easier to identify that a sustainable state is not present than it is to assert that it is present (Wilkinson and Cary, 2002).

From a narrow agricultural perspective, economic sustainability in headwater regions, in the context of land management, is likely to be expressed in the maintenance of the productive capacity of the land. In Bruntland's terms it is maintaining the productive capacity for future generations (World Commission on Environment and Development, 1987).

However, a broader approach to assessing the sustainability of agricultural production systems in headwater regions needs to consider both the adaptability of production systems, the adaptability of social systems as a consequence of wider economic systems, and the adaptability of land managers in responding to changed circumstances and to the failure of existing systems.

Assessing Sustainability

Sustainability in the broader sense is a concept like 'love', 'hope', or 'freedom': it tends to remain fuzzy until applied in a specific context (Government of Western Australia, 2002). Sustainability assessment can be process-based or ends-based. Pope, Annan dale and Morrison-Saunders (2004) contrast sustainability assessments which *identify and minimise adverse impacts* and *objectives-led sustainability assessment* which seeks to maximise sustainability objectives (such as increasing community and ecological sustainability). The theoretical ideal of sustainability as adaptability postulated by Boulding is a 'higher level' processed-based approach to sustainability.

In the extensive discussion and use of the term sustainable development since the World Commission on Environment and Development (1987) there has been recognition of three essential aspects of sustainable development. *Economic sustainability*, where an economy is able to produce goods and services on a continuing basis. *Environmental sustainability*, which maintains a stable resource base, maintains biodiversity and atmospheric stability, avoids over-exploitation of renewable resource systems, and where non-renewable resources are depleted only to the extent that investment is made in adequate substitutes. *Social sustainability* seeks to achieve fairness in distribution and opportunity, adequate provision of basic services including health and education, together with social accountability and social participation (Harris, 2003).

In managing land in a headwater region economic sustainability often will involve the ability of the land resource to produce livestock and crop goods and services for future generations. In any historical long-run consideration of economic 'development' the initial economic dependence on agriculture of a society in a given region is often replaced by increased reliance on non-agricultural activity (i.e. reduced agricultural dependency). Thus, assessing economic sustainability in both the short run and the longer run can be a difficult task.

Weak and Strong Sustainability

Considerable discussion of sustainable development in the economic literature has focused on the idea of *weak sustainability* and *strong sustainability* (Daly, 1994; Beckerman, 1994). The key difference between these two versions of sustainability is the capacity to make trade-offs between forms of capital (natural, human and social capital). The *weak sustainability* idea requires that the *total* value of man-made capital plus natural capital remains constant (or increases) over time. It allows an overall economic system to be sustainable by trading some forms of capital (human and natural) for others. A *strong sustainability* approach is based on the idea that substitutability between natural and man-made capital is limited – the two are seen as complements or factors that must be used together to be productive. The strong sustainability approach assumes that, because substitutability between natural and manufactured capital is limited, that stocks of natural capital, including fuels, minerals, soils, flora and fauna should not be depleted through time.

There has been considerable debate about the different positions represented by proponents of strong and weak sustainability. Strong sustainability has been defended by Daly (1999) and criticized by Beckerman (1994), who rejects the concept of sustainability in general (Harris, 2000). The significant issues are whether man-made and natural capital are substitutes or complements (Daly, 1999); and the extent to which natural capital is limiting. Costanza and Daly (1992) suggest that a minimum necessary condition for sustainability can be expressed in terms of the conservation of natural capital. This policy goal leads to two decision rules, one for renewable and the other for non-renewable resources. For renewables, the rule is to limit resource consumption to sustainable yield levels; for non-renewables the rule is to re-invest the proceeds from non-renewable resource exploitation into renewable natural capital. Following these two rules will maintain a constant stock of natural capital (Harris, 2003).

Broad and Narrow Social Sustainability

The tripartite approach to sustainability considers social sustainability to require, as crucial elements, the fulfilment of basic needs for health and education, and social participation. These seem to be minimum requirements which, in many societies, would need further refinement and extension to be suitable tests for social sustainability. Assessing social sustainability can be difficult because to assess whether a social 'state' is sustainable we have to agree on the desired social states to be sustained (the basic needs to be met), the level at which the desired state should be present, and over what time frame and what spread of social geography—the local community, the region, or larger entities. As a society becomes more affluent, the level of appropriate provision of 'basic need' will change.

Some of the above questions give rise to an issue that can be considered *broad* and *narrow* social sustainability. This is analogous, but different, to weak and strong sustainability.[1] Broad social sustainability implies a broader frame of reference—the region rather than the community, or the economy rather than the region—where labour and capital are mobile and responsive to changes in price signals; and new opportunities for social and community development might ensue. It suggests that socio-economic sustainability has some limitations as a concept at the small scale e.g. an individual town or community level and should be applied at higher levels of aggregation. Broad social sustainability highlights the need for dynamic adjustment process in the communities where individual towns and communities are part of larger social and economic processes.

Narrow social sustainability implies that existing social structures and networks should be preserved, if necessary, in spite of confounding economic and demographic pressures. While the maintenance of narrow social sustainability in discrete geographical locations is, at least in the short term, generally considered desirable (see, for example Pretty, 1995), it presents the potential possibility of communities being non-responsive to economic change—to the ultimate detriment of community sustainability. Where adaptability to changing external social and economic forces is not present, social sustainability is unlikely to be maintained other than by the provision of external government support. In responding to the pressures of change, it is important for communities to be able to distinguish between local forces for change which may be effectively resisted and broader societal forces for change to which communities need to adapt (Stayner et al., 2000).

Yenken and Wilkinson (2000) present a definition of social sustainability as the continuing capacity of a society to provide for the well-being of all its citizens and its ability to do so in a fair and equitable way. Such a definition embraces elements of both narrow and broad social sustainability. It implies that broader society has the responsibility to redress narrow social

[1] Stoneham et al. (2000) developed a similar analysis to the one that follows using the nomenclature of weak and strong social sustainability. The concepts of weak and strong sustainability are more usefully restricted to consideration of whether man-made and natural capital are substitutes or complements. The issues for social sustainability involve other unrelated concepts. While social development and social sustainability do involve social capital, the idea of social capital is still ill-defined (see for example Castle, 2002).

unsustainability. Stoneham et al. (2000) noted that Yenken and Wilkinson tend to define safe minimum standards as a relative rather than absolute concept.

Typical objectives for social sustainability at a headwater or regional level are likely to be couched in terms of narrow sustainability. For example, the minimisation of social disadvantage and social dislocation, ensuring and enhancing community vitality and maximising employment opportunities. Social dislocation is likely to occur where a significant proportion of the population relocates or changes its employment status. Social disadvantage and social dislocation occur as a consequence of the reduced capacity of a headwater region to support local communities. Such social 'states' are as likely to be a consequence of regional or national economic impacts as they are to be consequences of local ecological unsustainability.

A Framework for Analysis of Resilience in a Social-ecological System

A common way of testing the adaptability of socio-economic systems is to assess the current situation, identify the key processes and consequences of changes or perturbations to the constituent systems and grapple with the complexity of interactions of ecosystems, economic systems and social systems by establishing possible scenarios and assessing the resilience of these scenarios (Figure 1).

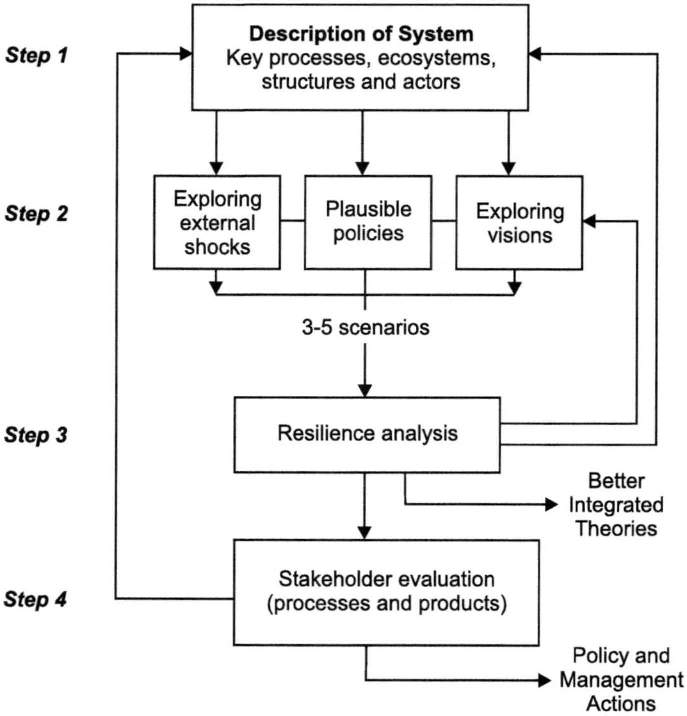

Figure 1: A framework for the analysis of resilience in social-ecological systems (Walker et al., 2002).

In the analysis of the sustainability of the upper catchment study areas in the Murray-Darling Basin presented below two broad scenarios were considered. One was that, with widespread adoption of management practices currently recommended for sustainability, existing land uses would achieve environmental, social and economic sustainability in the upper catchments. An alternative (group of) scenario(s) was that, where improvement in the management practices of existing use of land would not produce sustainable outcome, alternative land uses would be identified and assessed for sustainability.

Australian Goals for Sustainable Development

The nationally stated goals for sustainable development in Australia are expressed as 'using, conserving and enhancing the community's resources so that ecological processes on which life depends are maintained and the total quality of life, now and in the future, can be increased' (Australian Government, 1992). Within this objective there are significant community expectations that rural landscapes will provide:

- Clean water, and sufficient water flows and water reserves, for environmental, industrial, agricultural and urban use;
- Terrestrial and aquatic biodiversity;
- Profitable land-using enterprises which provide regional development, food production, and support resilient communities and resilient economies; and
- Resources that meet aesthetic, recreational and spiritual needs (Hajkowicz et al., 2003).

It has usually been difficult to meet all of these objectives satisfactorily.

Natural resource management is as much to do with people, what drives them and their capacity to respond, as it is to do with the biophysical. In this context, it is important to understand the social capacity and resilience of local communities to respond to change. It is equally important to know whether and how social capacity can be enhanced to assist landholders and local communities in responding to changed natural resource management regimes.

Sustainability in the Murray Darling Basin Catchment

The Murray-Darling Basin (MDB) is Australia's largest river catchment (approximately one million square kilometres) being the catchment for the Murray and Darling rivers and their many tributaries. The catchment extends from southern Queensland to South Australia and including three quarters of New South Wales and half of Victoria (Figure 2). The catchment has one exit flowing out of Lake Alexandrina in South Australia. The MDB generates about 40 percent of the national income derived from agriculture and grazing. The Basin contains more than twenty major rivers as well as important groundwater systems. It is also an important source of fresh water for domestic consumption, agricultural production and industry.

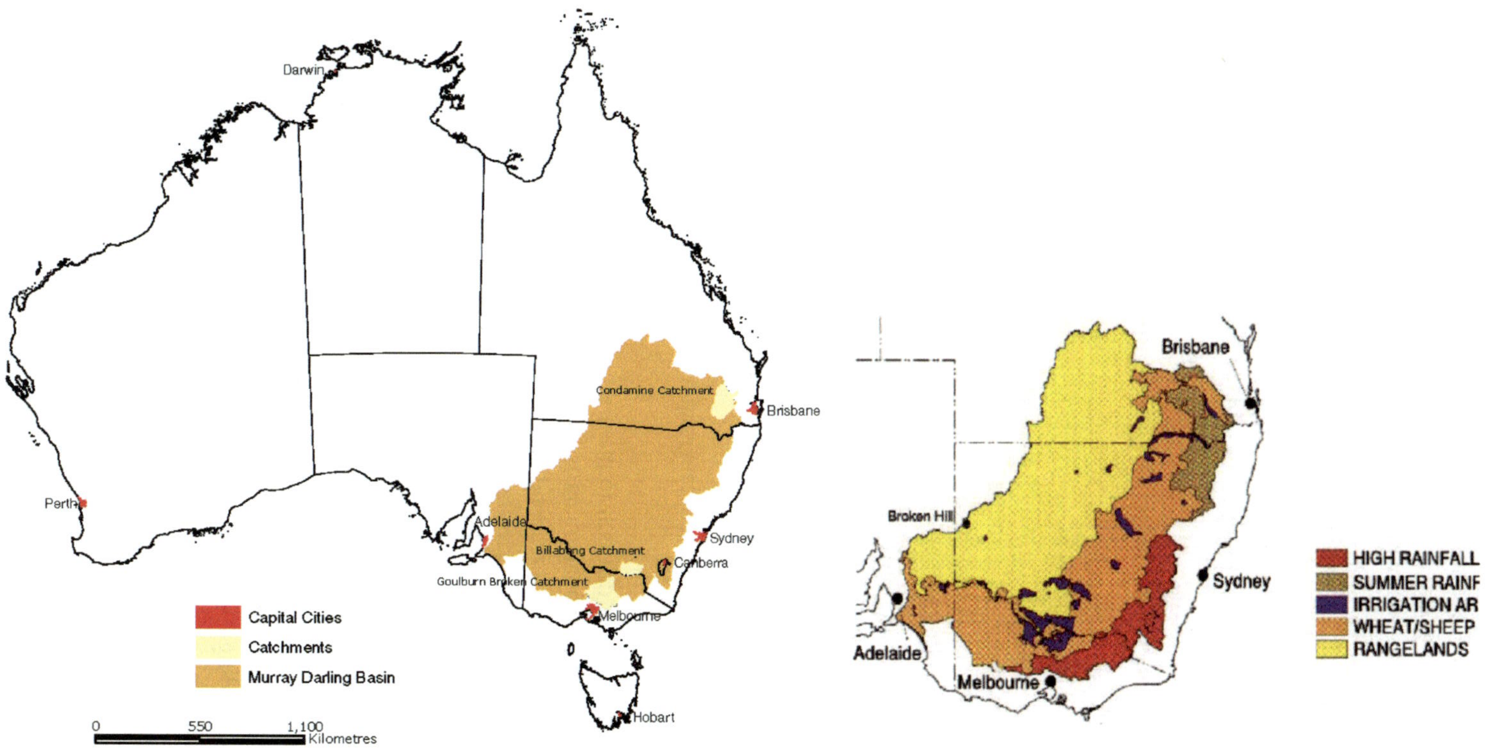

Figure 2: Murray-Darling Basin pilot study catchments and the basin's agricultural regions.

The MDB represents a major proportion of the productive agricultural land of Australia and embraces a variety of agricultural practice, ecological diversity and landscape form. Environmental degradation is an increasing problem and is particularly expressed in the degradation of the Basin's waterways. However, not all the land resource is unsustainable in its current agricultural use. In response to increasing evidence of environmental degradation, the Murray-Darling basin Commission initiated a four-year (1999-2003) research project (the Landmark project) to assess the sustainability of current dryland uses and farming practices in the upper reaches of the basin (Avery et al., 2005). The project was designed to identify needed land use, land management, and government policy change for improved sustainability in the upper reaches of the MDB.

The sustainability study was undertaken in three upper catchments of the MDB representing three major extensive land uses for the basin (Figure 2):

- high rainfall grazing (>600 mm pa, winter dominant) in the *upper Goulburn Broken Catchment* in Victoria;
- cropping and grazing (wheat/sheep) with winter dominant rainfall of the *upper Billabong Creek Catchment* in southern New South Wales; and
- summer rainfall dominant cropping landscapes in part of the *upper Condamine Catchment* of south-eastern Queensland.

The assessment of environmental sustainability focused on water balance, nutrient balance, terrestrial biodiversity and ecosystem health, and particularly on maintaining the *productive capacity* of soils and natural resources and the *stock of biological assets.* Environmental sustainability was assessed in terms of biophysical sustainability—soil erosion, soil acidity, soil nutrient management, and water management. For most of these criteria biophysical models were developed to assess sustainability under varying climatic and other situations. Relevant environmental indicators were applied at a pilot region scale and farm productivity was assessed at paddock scale.

Assessing Socio-economic Impacts

The assessment of socio-economic sustainability focused on the social consequences of any current declining land productivity, the economic consequences of different land management scenarios, and the demographic and socio-cultural consequences for relevant communities.

There are well-identified regional social and regional economic consequences of significant changes in land and land-use sustainability. Some of the relevant social and economic measures of socio-economic sustainability are shown in Table 1. These show minimisation adverse impacts on sustainability.

Table 1: Social and economic consequences of land-use change and rural adjustment

Economic Factors		*Social Factors*	
Regional agricultural production		Demographic change	• employment change • population migration • change in age profiles
Related regional goods and services		Services	• health and medical • education • retail services
Regional wages or income		Institutional vitality and social capital	• community networks • voluntary organisations • capacity of local leaders
Employment	• number • type • seasonality		
Infrastructure			• minimal social dislocation • public/private partnerships

(Adapted from McDonald, 1990)

The analysis of the current situation and time series changes with respect to social and economic sustainability was based on the factors identified in Table 1 where data were available. Good data sets were available for most measures other than institutional vitality and social capital. The data for developing regional social profiles was provided by the various data collections of the Australian Bureau of Statistics (ABS), particularly the ABS Census of Population and Housing and the Australian Agricultural Census. ABS data collected at the Statistical Local Area (SLA) was used for the analysis describing the current socio-economic state of the three upper catchments.

As a basis for assessing social impacts of current recommended practices and land-use change, a social profile for each of the three regions was developed. The profile described the current socio-economic structure of the region using an analysis of standard socio-demographic variables (Table 2). The social profile also described the trajectory of change in a region's agricultural community over the past 25 years. These social profiles provided a basis for understanding the capacity for each region to respond to change.

A model of demographic change in the farm population was developed based upon observed levels of entry and exit from farming (Barr, 2004), building on an earlier model of net changes in age cohort population by Clawson (1963). Using a simple stock and flow model, with 5-year age cohorts, entry and exit estimates were based upon migration data from the Census of Population and Housing.

Table 2: Examples of social profile indicators

Indicator	*Upper Goulburn Broken*	*Condamine*	*Upper Billabong Creek*	*Non-metro-politan Australia*	*Australia*
Total population, 2001	71,667	129,775	13,500	na	na
Population density, 2001 (people/km^2)	4.6	2.8	2.2	na	na
Median age, 2001 (years)	36	33	39	37	35
Farmer population, 2001	2,157	3,311	1,033	na	na
Change in farmer population 1996-2001 (%)	–1.3	–2.3	1.1	na	–2.1
Median age of farmers, 2001 (years)	55	50	51	50	50
Agricultural employment, 2001 (% of workforce)	9.4	8.7	21.6	8.9	3.5

Agricultural Dependency and Social Sustainability

The need for local farming land use practices to be ecologically sustainable in the long run is a necessary, but not sufficient, condition for socio-economic sustainability—particularly when considered in terms of 'narrow' regional social sustainability. The regional social and socio-economic impacts for rural headwater communities, as a consequence of either declining resource productivity or as a consequence of changing external markets for local agricultural products (e.g. T.E. Lawrence's camel markets referred to earlier) will be influenced considerably by the extent of a region's agricultural dependency.

Some regions or communities less dependant on agricultural resource use will be more robust to changes in land use and management practice and have greater potential for adaptability to a changing environment. Communities more dependant upon agriculture are more vulnerable to the impacts of adverse change. The social capacity to adapt to significant ecological or market change is likely to be positively assisted by non-agricultural resource endowments and to be inversely linked to regional agricultural dependency.

The three pilot regions had different level of agricultural dependency, reflected in indicators such as agricultural employment as a percentage of the workforce (Table 2) and the value of output of the agricultural sector compared to total regional output (Cary et al., 2004). Regional economic modelling, using input-output analysis, was used to establish the local contributions of relevant industry sectors to the regional economy and to model the economic

Table 3: Output, Gross Regional Product and Employment impacts of 5% reduction in productivity reflecting a reduction in land available across sheep, grains and beef cattle sub-sectors

	Upper Gouslburn Broken			*Upper Billabong Creek*			*Condamine*		
	Output (\$m)	*GRP (\$m)*	*Emp. (no.)*	*Output (\$m)*	*GRP (\$m)*	*Emp. (no.)*	*Output (\$m)*	*GRP (\$m)*	*Emp. (no.)*
Initial change	–6.333	–3.315	–66	–6.169	–3.255	–53	–11.631	–6.003	–90
Industrial effect	–2.324	–1.114	–19	–2.245	–1.059	–17	–5.203	–2.215	–32
Consumption effect	–1.671	–0.847	–16	–1.115	–0.556	–12	–2.499	–1.198	–21
Total effect	–10.328	–5.276	–101	–9.529	–4.870	–82	–19.333	–9.416	–143
% change from base	–0.5%	–0.5%	–0.5%	–1.9%	–2.0%	–2.0%	–1.1%	–1.2%	–1.3%

consequential changes in agricultural production as a result of declining agricultural productivity.[2] An economic assessment of the impacts on the value of output, gross regional product, and employment as a consequence of a 5% reduction in land available to broadacre agriculture for each of the catchments is presented in Table 3. There were considerable differences in the economic impacts between the catchments—particularly the upper Goulburn (low agricultural dependency) and upper Billabong (high agricultural dependency).

One of the catchments—the Goulburn-Broken—was in a state of significant socio-economic transition because it was within commuting distance (for weekend recreational farming) of the conurbation of Melbourne, with a population of approximately 3.5 million people. Thus, in the upper Goulburn catchment land values were very high in comparison to the value of traditional agricultural production able to be derived from this land. This catchment is in transition from an agricultural economy to a post-agricultural economy a transition driven by high landscape amenity and proximity to Melbourne. For most farmers, the strategy of purchasing land to achieve economies of scale is not financially feasible. Farm incomes were commonly supplemented with other sources of income and farmers were more likely to sell their land for 'hobby' farming. Agricultural landholders seeking to increase farm size frequently achieved this by purchasing farming land in other regions where land was more competitively priced. However, the new entrants with higher non-farm incomes reduce the region's agricultural dependency and increase the capital and social capacity to respond to land degradation reflecting an example of broad social sustainability. This catchment is thus undergoing considerable social and economic change. It is diversified and socially resilient with a declining role for 'production' agriculture and increasing role for leisure or 'lifestyle' agriculture providing a new basis for economic and social sustainability.

Traditional agricultural enterprises were a major source of employment in the upper Billabong catchment. There are many small farms and 50% of farms had gross incomes less than $100,000 in 2001. While there are some peri-urban effects of an adjacent provincial city, there are limited other industries and a high level of agricultural dependency. Many farmers in this catchment have limited ability to keep pace with declining terms of trade through increases in enterprise scale. Regional capacity for change has been less than for the other catchments and the options for agricultural diversification are limited. With slow rates of farm adjustment, and with limited expansion or subdivision options, this catchment is likely to be less economically and socially sustainable.

---

[2] The input-output analysis employed disaggregated National Accounts data to estimate the level of inputs 3 (labour, intermediate commodities, regional imports) to sectors of the regional economy and outputs (consumption, intermediate commodities, regional exports) from sectors of the regional economy (Cary et al., 2004).

The upper Billabong and the Condamine catchments would suffer more significant impacts (than the upper Goulburn) as a consequence of declining agricultural output due to land degradation (Table 3). In contrast, the Condamine catchment had a diversified and adaptable agricultural economy. However, in sub-regions of this catchment agriculture was specialised and more vulnerable to changing circumstances.

Conclusion

This assessment model was very useful in identifying important dimensions for appraising the likely consequences for social and economic sustainability in conjunction with assessments of ecological and resource sustainability. There was no single or dual set of variables which were sufficient as indicators of regional social capacity to adapt to changing land resource regimes. In assessing regional sensitivity to land use or management change a suite of social and economic indicators will be required. The following indicators or characteristics (not all of which have been discussed in this paper) were important in assessing social and economic impacts:

- level and distribution of regional income
- mix, diversity and scale of industry sectors
- nature and economic characteristics of existing rural industries and land uses
- age of farmers
- demographic projections of changing age profiles of farmers
- distribution of (economic) scale of property operations
- extent and scale of external economic and social impacts—particularly of adjacent large cities and conurbations on land ownership and future land use in the region
- level and distribution of property turnover and property subdivision
- extent of opportunities for off-farm income employment and income.

The strength of this assessment of social and economic impacts, within a broader social-ecological analysis, was that it employed secondary data for socio-economic analysis thereby confining costs to data purchase rather than the larger expense of data collection and thus avoided the logistical problems associated with primary data collection. The analysis incorporated the influence of adjacent towns and cities, outside the defined catchments, which often have a major social and economic impact in a given region.

A major strength of the approach was that it provided quantitative, spatially explicit data for appraising social and economic capacity to respond to ecological degradation and for assessing the resilience of social-ecological systems. This was an important step in informing and engaging communities, industry and governments in considering opportunities for improvements in land use and management. These groups can thus be informed of the scale of environmental and social impacts and the possible required changes to overcome these impacts. This is vital to make the link between catchment planning and on-ground action.

The search for more sustainable systems is an adaptive search for systems that are improvements on existing systems. The need for a *process* approach to improving our understanding of sustainability, and to transforming the way resources are used, emphasises the need to bring together research, community, and industry resources to better understand environmental, social, economic and legal systems and new technologies to encourage sustainability.

References

Anderies, 2003. Minimal models and agroecological policy at the regional scale: An application to salinity problems in southeastern Australia. Unpublished.

Australian Government, 1992. National Strategy for Ecologically Sustainable Development. Development of Environment and Heritage, Canberra, ACT.

Avery, A., Barr, N., Barson, B., Beverly, C., Cary, J., Clarke, D., Eigenraam, M., Gallant, J., Madden, B., McKenzie, N., Price, P., Ridley, A., Robins, L., Stoneham, G., Thomson, D. and Webb, T., 2005. Landmark: A method for testing dryland agricultural sustainability, Murray-Darling Basin Commission, Canberra ACT (forthcoming).

Barr, N. F., 2004. The Micro-Dynamics of Change in Australian Agriculture. Australian Bureau of Statistics, Canberra. ABS Catalogue No. 2055.0

Beckerman, Wilfred, 1994. Sustainable Development: Is it a Useful Concept? *Environmental Values,* **3:** 191-209.

Boulding, K.E., 1978. Ecodynamics. Sage. Beverly Hills, California.

World Commission on Environment and Development, 1987. Our Common Future. Oxford University Press, Oxford.

Cary, J. W., Webb, T., Barr, N.F., Charalambou, C. and Pinge, I., 2004. Social impact assessment to identify regional capacity to achieve and manage land use change. Murray-Darling Basin Commission, MDBC Publication 14/03. Canberra.

Castle, E.N., 2002. Social capital: An interdisciplinary concept. *Rural Sociology,* **67(3):** 331-349.

Clawson, 1963. Aging farmers and agricultural policy. *Journal of Farm Economics,* **45(1):** 13-30.

Costanza, R. and Daly, H.E., 1992. Natural capital and sustainable development. *Conservation Biology,* **6:** 37-46.

Daly, H.E., 1994. Operationalizing Sustainable Development by Investing in Natural Capital. *In:* Ann Mari Jansson et al. (Eds.), Investing in Natural Capital: The Ecological Economics Approach to Sustainability, Island Press. Washington, D.C.

Daly, H.E., 1999. Ecological Economics and the Ecology of Economics: Essays in Criticism. Edward Elgar. Northampton, Massachusetts.

Government of Western Australia 2002. Focus on the future: The Western Australia state sustainability strategy consultation draft. Western Australia, Department of Premier and Cabinet, Perth.

Hajkowicz, S., Hatton, T., Meyer, W., McColl, J. and Young, M. (2003). Conceptual Framework for planned landscape change. *In:* B.P. Wilson and A. Curtis (Eds.)

Agriculture for the Australian Environment. Johnstone Centre, Charles Sturt University, Albury.
Harris, J.M., 2003. Sustainability and Sustainable Development. Internet Encyclopaedia of Ecological Economics, International Society for Ecological Economics.
Holling, C.S. (ed.), 1978. Adaptive environmental assessment and management. John Wiley and Sons. London.
Lawrence, T.E., 1986. The Seven Pillars of Wisdom, New Orchard Editions, Poole Dorset.
Lee, K.N., 1993. Compass and gyroscope: integrating science and politics for the environment. Island Press. Washington, DC.
McDonald, G.T., 1990. Regional economic and social impact assessment. *Environmental Impact Assessment Review*, **10:** 25-36.
Pope, J., Annandale, D. and Morrison-Saunders, A., 2004. Conceptualising sustainability assessment. *Environmental Impact Assessment Review,* **24:** 595-616.
Pretty, J., 1995. Regenerating Agriculture: Policies and Practice for Sustainability and Self-reliance. Earthscan. London.
Stayner, R., 1997. The role of the regional economy in farm adjustment. Report to the Rural Industries R&D Corporation. RIRDC, Barton, ACT.
Stoneham, G., Eigenraam, M., Ridley, A. and Barr, N., 2003. The application of sustainability concepts to Australian agriculture: An overview. *Australian Journal of Experimental Agriculture,* **43:** 195-203.
Walker, B., Carpenter, S., Anderies, J., Abel, N., Cumming, G., Janssen, M., Lebel, L., Norberg, J., Peterson, G.D. and Pritchard, R., 2002. Resilience management in social-ecological systems: A working hypothesis for a participatory approach. *Conservation Ecology,* **6(1):** 14. [online] URL: http://www.consecol.org/vol6/iss1/art14
Wilkinson, R.L. and Cary, J.W., 2002. Sustainability as an Evolutionary Process. *International Journal of Sustainable Development,* **5(4):** 381-391.
Yenken, D.G. and Wilkinson, D., 2000. Resetting the Compass: Australia's Journey towards Sustainability. CSIRO Publishing, Melbourne.

4

Building Co-operations, Coalitions and Governance on Mountain Catchments Sustainability

Pier Carlo Zingari
European Observatory of Mountain Forests (EOMF)
350 Quai Charles Ravet, 73000 Chambéry, France

Introduction

This paper aims at providing some evidences that integrated and participatory watershed management (IPWM) is not only a theoretical framework but is currently stepping forward and becoming a necessary practice with successful examples. From international co-operations to local governance case studies, mountain areas are setting the stage of a new input to the sustainable management of natural resources.

At the launch of the International Year of Mountains at the UN, Professor Bruno Messerli emphasised some important new findings with regard to population and water resources in mountains: "mountains cover around 25% of the land surface and approximately 26% of the global population lives inside or very close to mountain areas."

Since the first international mountain conference of Munich, Germany, in 1974, we are using the never proved figure that 10% of the world population lives in the mountains. The new figure comes from a database at 0.5° resolution, derived from a 1 km gridded polygon file, that defined the spatial extent of 242 countries for which 1995 country-level population statistics were available (Meybeck et al., 2001). This means that mountains cannot be characterised only by remote, poor and disadvantaged people and communities, but also by urban centres inside and immediately outside mountain valleys as in the Alps, around the Pacific or even by mega-cities as in Latin America from Mexico City to Santiago de Chile.

In this sense, we have to rethink the highland-lowland interactive system in a new and global perspective. As perhaps even more important for humanity in the 21st century we have to consider the natural resources of mountain regions like water, biodiversity hotspots, recreation areas for an increasingly urbanised world, mineral resources, hydropower, agriculture, forestry etc.

- In the arid and semi-arid regions of the tropics and subtropics, where possibly more than half of the world population is living, 80 to 100% of the available freshwater is coming from mountain areas. Surface and groundwater in the Aral basin, the Nile, Indus and Euphrates, Jordan, Yellow River etc. are coming from mountains and uplands. The evaporation in the lowlands can be much higher than the rare precipitations. This means that the water balance in the lowlands can be negative, even without the deviation of water for households, industries and irrigation.
- A statement from Dutch scientists is astonishing that in the dry summer of 1976, 95% of the Rhine water flowing into the North Sea came from the Alps, from melting snow and ice at high altitude. But also in California, scientists are estimating that two thirds of the available freshwater for this highly industrialised state are coming from the winter and spring snow accumulation in the Sierra Nevada.

All this means that mountains as water towers play a fundamental role for water supply and food security. If the prediction is correct that we shall face a serious water crisis in the middle of the 21st century, then the mountains will play an absolutely central role, not only for the quantity and quality of water, but also for the conflicts about water.

Coming to Europe, in 2004 mountain areas surface and population involved was also reviewed by the European Union (Nordregio, 2004). Figures coming from a more detailed analysis of coupled GIS and statistical information provide now a realistic picture of the situation: mountains cover 41% of the land surface in the EU25, and approximately 20% of the total population.

Until 2004, no official figures were available for European mountain areas and only agricultural land was taking into consideration in a quite narrow perspective. The new figures imply now a broader view and perspective over mountain areas and water resources, as well as sectors and people involved. The latest, for example, are not only people living in the mountains but are the estimated half billion tourists enjoying European mountains annually.

The upstream-downstream relationships have been one of the key themes of both the International Year of Mountains, 2002, and the International Year of Fresh Water, 2003. It is clear that water and watershed are closely related aspects of these relationships and of the work of a relevant number of individuals and organisations since then.

In particular, and among others, a few examples can be selected to provide an overview of the main activities that are improving co-operation, allowing a better consideration of hydrology, ecology and water resources in mountain areas.

Some Examples of Co-operation, Coalition and Governance

One global example is the International Partnership for Sustainable Development in Mountain Regions or, more shortly, the "Mountain Partnership". This is a voluntary alliance of partners dedicated to improving the livelihoods of mountain people and the protection of their environments launched at the Johannesburg's World Summit for Sustainable Development in 2002. An MP secretariat is hosted by UN-FAO with the financial support from the governments of Italy and Switzerland. By the beginning of 2005, some 115 members had joined the Mountain Partnership, comprising countries, intergovernmental and non-governmental organisations.

Watershed management is one of the thematic areas of actions of the MP with a view to bridge gaps in knowledge, skills and expertise, to identify opportunities for collaboration and to connect those who requires assistance with those who can provide it. The work done can be normative (devising technical guidelines, methodologies and tools for watershed management and conservation), or field-related (assisting countries and communities tackling issues through capacity building, institutional strengthening and pilot initiatives).

While the Mountain Partnership has been developing, other areas of work such as watershed has been a central issue since the first meetings and decisions. One of the reasons is the relevant work of review done by FAO, mainly by its Forestry Department, from 2001 in the perspective of shaping a new generation of watershed management programmes and projects.

FAO started a comprehensive review of watershed management initiatives world wide with key international actors in 2001 raising some basic questions, such as:

- are we sharing experiences and lessons learned?
- are we using the appropriate participatory process?
- what kind of participation is taking place?
- to what extent can the participatory approaches be used?
- are we overestimating what can be achieved through participatory approaches?
- are the developed technologies producing the expected results?
- are project activities sustainable and replicable?
- to what extent have the institutional/organisational and legislative arrangements been developed?
- are the expected policies/strategies in place?

The review has been made through a large consultation based on, among others, a key actor survey, four regional workshops (France, Nepal, Kenya, Peru), an *ad hoc* Conference in Italy, and several case studies across the world. This impressive work is starting to be available in a series of analytical publications.

Table 1: Paradigm shift recommended for future watershed management (from FAO, 2005)

Present scenario	*Future scenario*
1. Treating the *symptoms* of watershed degradation (deforestation, soil erosion, siltation, decreasing production, etc.).	1. Identifying and treating the underlying *causes* of watershed degradation (e.g. lack of knowledge, poverty, population increase, demand for resources, improper land use).
2. Priority focus on *off-site/downstream costs* and benefits of WM, e.g. downstream infrastructure risk, decrease in floods and sedimentation, increase in water quantity and improved water quality for downstream users.	2. At minimum, equal priority to *on-site costs and benefits of watershed management*, e.g. improving and maintaining upland agriculture, forestry, and rangeland productivity, water quantity and quality and flow regimen.
3. Programme/project design that : (1) *overestimates government capacity* (2) *lacks adequate institutional/ organizational* arrangements (3) assumes *policy changes will occur*.	3. Programme/project design that provides for *adequate government capacity, institutional arrangements and assurance of policy changes.*
4. *Short term* planning and financing.	4. *Long term* planning and financing.
5. *Top down research and development* and transfer of technology to local stakeholders that is driven by donors and education and research institutions.	5. Emphasis on *stakeholder participatory learning* and technology development process that builds on indigenous technologies and addresses local research needs.
6. *Sector based* research and education.	6. *Integrated* and multi-disciplinary research and education.
7. *Encroachment* of integrated rural development approach with multi-sector steering committees and line agencies (which for the most part has been a failure) into the integrated watershed management concept.	7. *Multiple use management of natural resources* (renewable and non-renewable) with emphasis on water and soil resources in upland watersheds with development responsibility given to the relevant line agency with clear definition as watershed management interventions. IRD implemented on watershed scale should be separate from WM activities with both being part of a watershed development programme.
8. To some extent, policies relevant to WM have been based on *intuition and common myths*.	8. Policy should be based on *scientific and tested evidence.*

(Contd.)

Table 1 (*Contd.*)

9. Some measures for sloping lands are *designed to alter/modify the natural system* causing, in some cases, negative impacts on the system.	9. Methods should *be compatible with the natural system* and cause minimal if any change in the system.
10. The *top-down approach*, which was prevalent during the 1970-1980's, gave way to the "grass roots" *bottom-up approach*. However, it appears that neither of the extremes is the recipe for success.	10. Program and project design that *includes bottom-up planning and selection* of WSM interventions to ensure local level economic and social feasibility *with top-down* (screening) of those interventions to ensure technical and policy consistency.
11. WM sector is often given the responsibility for *nation-wide mandate for soil and water* conservation.	11. Require that *development sectors whose interventions have potential to adversely impact soil and water resources* (e.g. roads, forestry, range management, livestock, agriculture, mining and tourism) to be responsible for mitigation of adverse impacts on all natural resources.
12. Conservation management of the water resource in *upland (headwater) areas has been given low priority* on global scale.	12. In view of perceived increased demand for clean water and water scarcity becoming a major problem during the next decade [as identified in the FAO review] future WSM programmes/projects need to place major *emphasis on the water resource in upland environments.*

One of the key underlying shifts required is moving from the "blue water", or surface and runoff to "green water", managing water that is found in soil and vegetation (FAO, 2004). A number of countries recognise the need of placing more emphasis on ecosystems and more specifically on forest ecosystems and support international activities that may improve the understanding and planning of forest and water schemes (FAO, 2002).

In this direction, a complementary work is in progress on the aspects related to mountain ecosystems and forests in particular. In 2003, Shiga, Japan, and in 2003, Chambéry, France, two major events took place on forests and water showing the need to consider the relationships between vegetation, climate and soil in an integrated and cross-sectoral way, establishing the total economic value of resources, putting in place appropriate incentives and collaborative arrangements among all stakeholders. There is a need for more suitable and available data on the ecological, technical and economic components of these relationships (Ministry Agriculture of Japan, 2002; EOMF, 2004).

UNESCO stresses the forest and water issues referring to "forested catchments, especially in headwaters, as the principal sources of high-quality water for human and ecological needs in up- and downstream areas". Nevertheless "baseline data are lacking on the hydrological functions of headwater forested mountainous areas, notably in the humid tropics, upon which to base management decisions for allocating basin water resources". Such information is also required to assess scientifically hydrological impacts of global change (Bonell, 2004).

Water-related ecosystems, namely forests and wetlands, their role and possible approaches are at the centre of the activities of the Convention on the Protection and Use of Transboundary Watercourses and International Lakes under the responsibility of the United Nations Economic Commission for Europe. The aim is to provide a platform for government officials to meet experts from international organisations, NGOs and private sector, to share knowledge and the role of forests and wetlands in the water cycles and the advantages related to their sustainable use, protection and restoration, including economic and financial dimensions (UNECE, 2004).

A consistent collaborative step forward in the demonstration of assets and challenges coming from integrated and participatory approaches to mountain watershed management is the establishment of permanent pilot sites. A pilot site should serve, at a large scale, the international co-operation through comparative and replicable approaches, and, at the local scale, through an open supportive exchange and dialogue with the direct players.

In this sense, one of the international mechanism providing a coherent framework is the UNESCO HELP (Hydrology for the Environment, Life and Policies) programme developed within the activities of the International

A view on the Mont Blanc and UNESCO HELP Pilot Site (Photo SEA74)

Hydrological Programme of UNESCO. The HELP initiative is establishing a global network of catchments to improve the links between hydrology and the needs of society via the following activities:

- promoting stakeholder dialogues;
- defining the interface between water policy and law with science;
- identifying basins 'vulnerable' to global change;
- integrating forest and water issues;
- determining how to address food and water policy issues;
- achieving a more integrated, scientific approach;
- addressing scientific questions where a scientific infrastructure is lacking.

Among the 69 basin pilot sites under the world wide 2003-2004 scheme, 21 are found in Europe. The operational pilot site called "Pays de Savoie – Mont Blanc – Léman" is one of the mountainous one.

The relevance of this site is the strong links between the concept of IPWM, the specific framework provided by UNESCO and the field practices of local stakeholders. A broad international co-operation has been allowing a local dynamics towards an operational governance on IWRM. The structure and function of the site is based on the following aspects:

- upstream (Mont Blanc) and downstream (Lac Léman) – both transboundary – environmental and socio-economic interactions;
- five local groups involving stakeholders and priority water-related issues (natural hazards, water uses, forests and grasslands, agriculture, wetlands, tourism, land tenure, pollution, conflicts);
- regular meetings, exchange of information, development of indicators, monitoring, reporting, communication through a "Water in Mountains" website;
- involvement of institutions and programmes (universities, administration, EU programmes, protected areas) in a permanent exchange platform;
- organisation of international conferences "Water in Mountains" every four years to review progress and exchange experiences.

Concluding Remark

Experiences and examples of IPWM show that progress in building co-operation, coalitions and governance is taking place at the international and at the other level, including the local one. There is still much to do, especially in environmentally and socio-economic fragile areas like mountains. Lessons learned from specific initiatives, such as the FAO watershed management review, the UNECE water and natural resources activities and the UNESCO HELP network of permanent pilot sites, are the best opportunities to transfer the essential of the progress done by a large set of individuals, institutions and organisations down to the diversity and complexity of the local situations.

References

Bonell, M., 2004. How do we move from ideas to action? The role of the HELP Programme. *Water Resources Development,* **20(3)**: 283-296.

EOMF, 2004. Proceedings of the International Workshop on Forests and Water, Chambéry, France, 5-6 June 2003, EOMF, Chambéry, pp. 50.

FAO, 2005. The Next Generation of Watershed Management Programmes. FAO, Rome.

FAO, 2004. Report of the Twenty-Fourth Session of the Working Party on the Management of Mountain Watersheds, Crakow, Poland, 9-11 September 2004, FAO/EFC, Rome, pp. 18.

FAO, 2002. Report of the Sixteenth Committee on Forestry. FAO, Rome.

Messerli, B., 2001. Contribution to the Official Launch of the International Year of Mountains IYM 2002, New York, 11 December 2001 (http://www.mtnforum.org).

Ministry Agriculture of Japan, 2002. International Expert Meeting on Forests and Water, Shiga, Japan, 20-22 November 2002, Tokyo, pp. 361.

Nordregio, 2004. Mountain Areas in Europe. Analysis of Mountain Areas in EU Member States, Acceding and Other European Countries. EU DG Regional Policy, Nordregio, Stockholm, pp. 271. (http://europa.eu.int/comm/regional_policy/sources/docgener/studies/studyen.htm).

UNECE 2004. Report of the Third Meeting of the Parties, Work Plan for the Period 2004-2006. Geneva, pp. 17.

5

Developing Sustainability Priorities with a Participatory Process: Lake Victoria Basin, East Africa

Leif Lillehammer, Terje Kleven[1], Tore Hagen, Mark Bain[2] and David Lewis[3]
Sweco Grøner, Lysaker, Norway
[1]Norwegian Institute of Urban and Regional Research, Gaustadalléen 21 N-0349, Oslo, Norway
[2]Natural Resources, Cornell University, Ithaca, New York 14853, USA
[3]Cornell Institute for Public Affairs, USA

Introduction

Our study headwater ecosystem is the Lake Victoria basin of East Africa. This ecosystem holds world leading status for freshwater lake size, vertebrate diversity elaboration, species extinctions, exotic species invasions, and freshwater fishery production. It is high in elevation, the source of the Nile River, situated at the centre of the tropics, mostly enclosed by highlands and mountain ranges, and nearly a closed hydrologic system. The ecosystem is old but erratic in persistence on the scale of evolutionary time. Humans may have emerged in the basin, persisted there throughout human history, and recently increased to substantial density. However, the region remains undeveloped with meagre means of transportation, energy production, and industrial activity. Finally, this ecosystem is presently receiving heavy international development assistance including one of the largest and most costly ecosystem study programmes on earth. Lake Victoria has displayed massive ecosystem change in the relatively short three decade period thought to span an original intact system to one still foundering in unanticipated ways (Kaufman, 1992; Goldschmidt et al., 1993; Goldschmidt, 1996; Verschuren et al., 2002).

Joint management of Lake Victoria Basin as a shared ecosystem can be traced back to the recommendations in Agenda 21, following the Rio Summit of 1992. Various discussions took place after this, culminating in the establishment of the Lake Victoria Environmental Management Project (LVEMP) through a Tripartite Agreement (Kenya, Uganda and Tanzania) signed in 1994. This was closely followed by the signing of the Lake Victoria Fisheries Organization (LVFO) Convention in 1996. At the same time revival of the East African Community (EAC) took place in 1996 and various activities has been undertaken through this body, e.g. tripartite agreements, protocols, treaties, programmes and projects.

Specific visions have been elaborated for these different projects, programmes and institutions (e.g. LVFO, LVEMP), but no comprehensive, overall and agreed upon vision for sustainable development of the lake basin resources were in place. The LVEMP Regional Policy Steering Committee therefore in May 2001 decided that EAC should spearhead the vision and strategy development for Lake Victoria and its basin. This was later mandated by the Committee on Lake Victoria Development Programme (CLVDP). Thus a project was launched, termed The Vision and Strategy Framework for Management and Development of Lake Victoria Basin. It aimed at being implemented in recognition of the need to create harmony and reduce conflict over management of the lake basin resources, and to eliminate future duplication and overlaps between various programmes/projects. Concurrently, National Task Forces (NTF) were established in each of the three countries to spearhead the national processes. A Regional Task Force (RTF) was also established at EAC level to review and harmonise the national processes to finally reach at a regional Shared Vision and agreed upon strategic framework. The project has been undertaken under the general supervision of the EAC, and it was facilitated by scientists and consultants from Norway, USA and East Africa. The NTFs and the RTF as well as the various Lake Victoria stakeholders, who took part in the process, are the owners of the Visions and the strategic framework and the ones who defined them.

Study Area and Context

Being one of the mightiest headwaters in the world, with a basin size at 190,000 km^2, and a surface area of 68,800 km^2, Lake Victoria drains into the White Nile in Jinja, Uganda. The size of the lake makes it vital for weather and climate modulation in the region. The lake basin, with all its tributaries rests in the Great Rift Valley, where historical and cultural traits can be found back to the dawn of humankind. The lake itself was first formed 750,000 years ago, but was completely dry 12,400 to 14,000 years ago (Johnson et al., 1996). Most (82%) of the water entering the lake comes from precipitation and a lesser part from stream flow and basin runoff (Bootsma and Hecky, 1993; COWI, 2002). This is balanced by a large rate of evaporation and a relatively small discharge to the White Nile (mean annual discharge 1046 m^3/s).

Within the Lake basin area is Tanzania occupying 44%, Kenya 22%, Uganda 16%, Burundi 7% and Rwanda 11%. Lake Victoria Basin has a population of about 30 million people of which approximately 25 million live in the three riparian states (Ntiba et al., 2001). The Lake basin is endowed with rich natural resources and provides food and water for domestic, agricultural and industrial use. More than 80% of the population in the Lake basin are engaged in agricultural production (Sanchez, 2002). The fish resources of the lake sustain, directly or indirectly, livelihood for about three million people engaged in subsistence, artisanal and commercial fishing. The fisheries are very important as a source for foreign exchange earnings with an annual landed value of 300-400 million USD (Abila and Jansen, 1997). At the same time the lake is the final recipient of human and industrial wastes and eroded soils from natural and human-initiated processes in the basin. The multiple activities in the Lake basin have increasingly come into conflict due to several negative trends and driving forces, often working in combination.

In general, stakeholders in the three countries have recognised the fragile lake ecosystem and the constraints for development but also pointed to the immense potential of the lake basin as an important social and economic resource that is not fully utilised.

Since the Lake Victoria Basin is one of the major headwaters of the downstream Nile, future development and utilization of its resources have the potential of affecting downstream riparian states. Thus we will shortly visit some important issues related to this.

In transboundary river basins usage of the water resources by upstream and downstream states is the potential bearer of conflicts (or water wars), as seen so frequently also in the middle east (Bulloch and Darwish, 2003; Shiva, 2002). The usage of Lake Victoria and its effect on downstream riparian states is but one example. Of the ten countries sharing the river Nile basin[1], five share the Lake Victoria headwater or sub-basin. Although the White Nile branch supplies less water to the downstream Nile than the Blue Nile[2] and water resources activities has overall less impact downstream, there has been a rising tension over the usage of the Lake Victoria, its riparian states and downstream Sudan and Egypt.

A major share of origin of this tension rests in the Nile Waters Agreement (NWA)[3], of 1929, and the allocation of its water between Egypt and Great Britain (which represented Uganda, Kenya, Tanganyika (now Tanzania) and Sudan). The treaty stipulates that no work of any kind may be undertaken on the Nile, its tributaries, or on the lakes which form its course, without Egypt's consent; and in particular if these works are related to irrigation or power generation, or if they affect the volume of water which reach Egypt or in any other way be detrimental to it (Nicol, 2003). It also gave Egypt the right to

---

[1] Burundi, Democratic Republic of Congo, Egypt Eritrea, Ethiopia, Kenya, Rwanda, Sudan, Tanzania and Uganda.

[2] 30% versus 60% as measured at Aswan according to Nicol (2003).

[3] The tension was furthered by a new agreement in 1956 between Sudan and Egypt, increasing water allocation to themselves.

"inspect or investigate" the whole length of the Nile to the remote headwaters of the Nile (Raphaeli, 2004).

Although the recent Nile Basin Initiative (NBI) was launched to counteract the potentials for such conflict, some recent clashes over the White Nile waters have still surfaced. This includes launching of a Tanzanian project in 2004 to draw water from Lake Victoria and Kenya's "treat" of withdrawal from the NWA in 2003 (Raphaeli, 2004). Nevertheless, the NBI is moving forward and has a process-oriented approach. It firstly sought for a common point of departure for all stakeholders – the NBI Shared Vision (NBI, 2001).

"To achieve sustainable socio-economic development through equitable utilization of, and benefit from, the common Nile Basin water resources."

The NBI Shared Vision thus framed the starting point for subsidiary action programmes (SAP) at sub-basin level, and one major objective was to pursue cooperative activities and not conflicts. As with the rest of the Nile riparian states the five Lake Victorian riparian states has embraced the NBI Shared Vision and the relevant SAPs[4]. Thus, a Shared Vision and agreements upon sustainable development of the resources in the Lake Victoria Basin is relevant for the NBI.

Methods

The overall objective of the Lake Victoria Vision process was to develop broad consensus on realistic, achievable visions and strategy frameworks for long-term sustainable management of the Lake Victoria Basin. The process was intensively participatory. Over 12,000 stakeholders from community, district, national, and up to regional level, defined and harmonized the desired futures (visions) and the roadmap (strategies) to reach there. The process ended up in sectoral, national and regional visions and strategy frameworks, and indicators to monitor the further process. The visions and strategies cover multiple aspects of basin management, however, with water resources as a core. The formulation of shared visions and strategies was based on a set of overall values and principles guiding the process. This ended up in definition of five broad and mutually interlinked policy areas:

- *Ecosystems, Natural Resources and Environment* (resource management, protection and conservation);
- *Production and Income Generation* (natural resources utilisation);
- *Living Conditions and Quality of Life* (income and poverty, social services, health and education);
- *Population and Demography* (population development and migration);
- *Governance, Institutions and Policies* (development and harmonisation of policies, institutions and implementation machinery).

[4] One of these being the Nile Equatorial Subsidiary Action Programme (NELSAP), with headquarters in Kigali, Rwanda.

Each of these policy areas, or fields of activity, may be regarded as a driving force for change. However at the same time they are recipient of impacts from changes taking place in the other areas. Healthy ecosystems, that provide the population with goods and services (see for example, Costanza et al., 1998), are needed to reduce poverty. Vice versa, reduced social poverty reduces the strains on the environment. Sound environmental management is an integral part of the conditions for sustainable economic development; sound resource management sustains the functioning of ecosystems and their production of goods and services to the population.

In each of the three East African Community member states (Uganda, Kenya and Tanzania) where the exercise was undertaken a common approach was developed. Workshops were first arranged at community/village/ward level (hereinafter termed community level), then subsequently up to district and national levels and finally at regional (East Africa Community) level. Workshops at community level were the major activity of stakeholder interaction. In the sample localities selected workshops were held with local stakeholder representatives, village leaders and officers, religious leaders, women groups and civil society organisations. A series of participatory techniques and approaches were used to facilitate discussions of the present situation, visions, priorities and strategies[5]. With local variations, the workshops normally progressed in the following steps: (*i*) Description of the present situation; (*ii*) Formulation of visions and harmonisation of them across stakeholder groups; (*iii*) Formulation of prioritised strategies needed to achieve the vision and; (iv) Identification of indicators to monitor and evaluate change.

A "harmonisation matrix" was used to capture, summarise and report stakeholder opinions. The matrix defined a common framework and checklist to sum up stakeholders' conclusions. It included the five policy areas and (*i*) threats, (*ii*) priorities, (*iii*) strategies and (*iv*) indicators within these. District workshops (with selected representatives from the village workshops) were undertaken with the common aim to harmonise inputs from community level workshops within each district.. Using the matrix format, the district workshops summed up inputs from the local level and prepared material for the subsequent national workshop. National level workshops were held in all three countries to harmonize the visions at national level. The workshops were attended by selected participants from the village and district level processes as well as representatives from the private sector, focal ministries, national environmental and land use agencies, NGOs and civil society representatives and the media. The outcomes of the national workshops have been presented in the three National Reports (Statkraft Grøner et al., 2003). Table 1 summarises some data on the activities of the NTFs:

---

[5] Some of the specific tools used were "Historical Transect Analysis", "Mobility maps", "Livelihood Analysis" and different (flip chart) methods for setting priorities.

Table 1: Villages/local communities visited (workshops held) during the national processes

	Kenya	*Tanzania*	*Uganda*
Number of villages/local communities visited (workshops held):	66	31	10
Name of districts where the villages are located	Kisumu, Bondo, Kakamega, Busia, Trans-Nzoia, Suba, Kericho, Gucha, Migori, Nyando, Rachuonyo	Mwanza Reg: Mwanza city, Misungwi, Geita, Magu; Kagera Reg: Bukoba urban, Bukoba rural, Muleba (mainl. and island); Mara Region: Musoma urban Serengeti, Tarime; Bunda	Kampala, Bugiri, Rakai, Mayuge Mubende
Approximate number of participants in village workshops (average/total)	130/8000	50/1500	185/1865
Number of Lake basin districts represented at the district workshops (no/total)	11/29	18/18	5+/17
Approximate number of participants in district workshops	330	120 (40x3)	200
Approximate number of participants in the national workshop	60	95	85

The RTF regularly meet during the process to coordinate the NTFs and the country processes. A final regional workshop with broad representation from the RTF, NTFs and additional stakeholders was held to harmonize visions, priorities and strategies across the region based on the national inputs. This included also harmonisation and definition of indicators. The outcome, comprised a harmonised overall regional vision, sub-visions for the five policy areas as well as agreed upon strategies and indicators of change for each of the five policy areas. The whole process is summarized in Figure 1.

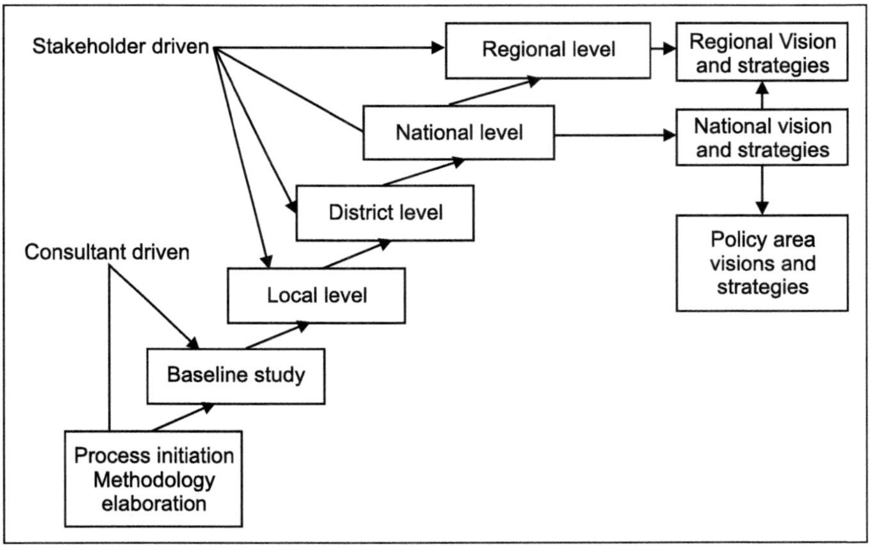

Figure 1: The overall process of the project with its major outputs.

Results

The National Visions

The agreed national visions are phrased differently, but encapsulate some common, shared principles and values. The national visions (see Box 1) express that future development in the Lake basin will have to take into consideration three closely interlinked policy issues or dimensions:

- Economic development as a basis for prosperous populations and improved livelihoods;
- Sustainable use of the natural resource base; and
- Protection of the environment.

Box 1: The agreed upon separate National Visions for Kenya, Uganda and Tanzania

Kenya: *"A Lake basin community enjoying a sustained natural resource base within a well conserved environment providing foundation for economic vibrancy that improves their livelihoods and increases opportunity for full realization of their potential".*

Tanzania: *"A Basin with the resources sustainably managed and communities having high standards of living".*

Uganda: *"A Lake Victoria Basin which is well planned and managed providing sustainable benefits, with a prosperous population meeting their development needs without depleting the natural resources and degrading the environment"*

The Regional Vision

The agreed Regional Vision includes all dimensions inherent in the national visions but is at the same time formulated in a way that distinguish the regional vision from those of the partner states. The Regional Vision for Lake Victoria Basin reads as Box 2.

Box 2: The Regional Shared Vision for the Lake Victoria Basin

"A prosperous population living in a healthy and sustainably managed environment providing equitable opportunities and benefits."

Note the close correlation with the Shared Vision of the NBI[6]. A prosperous population is closely linked to "sustainable socio-economic development" as phrased in the NBI and both Shared Visions emphasize equitable utilization/ opportunities and benefits.

Regional Visions for the Policy Areas

The national processes have also been the primary basis required to reach harmonized *regional visions and prioritized strategies* for the five policy areas. The agreed upon policy area visions are reproduced in Box 3.

Box 3: The agreed upon policy area visions

1. **Ecosystems, Natural Resources and Environment:** *"A prosperous livelihood and enhanced management of ecosystems, natural resources [and] a clean and healthy environment."*
2. **Production and Income Generation:** *"Resources sustainably and equitably utilised for increased income and poverty reduction in the Lake basin."*

[6] The NBI Shared Vision was developed during a 6 month period (Nicol, 2003), whereas the Lake Victoria Shared Vision was developed during an eighteen months period.

3. **Living Conditions, Poverty and Quality of Life:** *"A healthy, well educated society with high quality of life, well developed infrastructure and free from poverty."*
4. **Population and Demography:** *"A healthy, competent and productive population able to utilise and manage natural resources sustainably to achieve economic growth and development."*
5. **Governance, Institutions and Policies:** *"An empowered and gender sensitive community that observes the rule of law and human rights, well integrated institutional framework enabled by a policy environment that facilitates their involvement in the management of resources".*

A wide array of strategies and indicators was identified during the national processes. Further harmonisation and prioritisation was conducted in the regional discussions. Prioritised strategies and indicators were identified and elaborated for each of the five policy areas to reach the goals expressed by regional policy area visions. The purpose was to decide on the most critical issues to be the object of further strategy development. These are however too extensive and numerous to go through in this article however selected harmonized strategies and indicators within the policy area Ecology, Natural Resources and Environment are shown in Table 2.

Table 2: Prioritised strategies for Policy Area 1 - Ecology, Natural Resources and Environment

Sector strategies	*Indicators*
1. *Fish resources management*	
• Promotion of sustainable fisheries policies, baseline surveys, extension services, monitoring, surveillance and control measures	Maintained fisheries yield at an estimated long-term level Reduced numbers of licences and quotas
2. *Land use and natural resources management*	
• Improve land use and natural resources planning with particular focus on urban expansion	Decrease in encroachment into vulnerable areas. Improved land cover
• Promotion of proper land use management practices	Number of approved land-use plans in place
• Promotion of co-management of wildlife	Number of wildlife co-management areas established
• Wildlife inventories conducted	Increase in wildlife inventories
• Promote the establishment of community forests and woodlots/ afforestation/tree planting schemes/agroforestry	Forest cover/number of trees

(*Contd.*)

Table 2 (*Contd.*)

3. *Water resources management*	
• Promotion of integrated water resource/water catchment management	Reduced sediment loading into the Lake
• Promotion of water quality and quantity monitoring	Improved water quality
• Intensify water weed monitoring and control	Reduced density of weeds
4. *Pollution control and waste management*	
• Integrated waste management of solid wastes and discharges into the Lake	Reduced area of lake experiencing anoxia at any depth
• Reduction of point source pollution by strengthening industrial and municipal waste management in urban and industrial centres	Increased tonnage of human and animal waste properly disposed and treated
• Regulations to reduce and control pollution from mines	
• Regulations and promotion of farming methods to reduce un-controlled use of pesticides, herbicides and fertilizer	

Discussion

Although the agreed upon Regional Vision and the subsequent reporting finalization marked the end of the involvement of the International Professionals in the process, the process is still ongoing. Further, the Council of Ministers (for Kenya, Uganda and Tanzania) at their 7th Meeting in January 2004 adopted the Regional Vision, and feedback and implementation of the various harmonized visions and strategies are currently taking place at national, district and community level.

Since the three partner states of East African Community have reached at a common platform and developed harmonized shared vision(s) and strategies for sustainable development of the lake Victoria basin[7], it implies cooperation in further action within developing the common resources. This might have further implication on their dialogue and cooperation with the downstream Nile states, e.g. Sudan and Nile. Furthermore, the Shared Regional Vision and its various follow up actions are also a tool for reducing conflict between the Lake Victoria states themselves over their joint recourses, leaving them also in a more joint and harmonized position when discussing about utilization of

[7] And it has been adopted at Ministerial level.

the Lake Vizctoria headwater resources with the downstream Nile states. Lastly, the Regional Shared Vision of Lake Victoria basin and the NBI Shared Vision bear similarities and are a platform for further collaboration and dialogue related to utilization of resources of this major headwater of the Nile. If some of the bottlenecks and conflicts can be overcome we firmly believe that if the stakeholders of the Lake Victoria basin can connect the visions to actions, this will have the potential of positively influencing the future development of the lake basin. Nelson Mandela has phrased this interconnection in a clear-cut way: "A vision without action is just a dream; an action without vision just passes time; a vision with an action changes the world".

Acknowledgements

The contributors to the project and process are innumerable. However, most importantly thanks go to all the enthusiastic Lake Victoria stakeholders that participated in the process in developing visions and strategies at various levels. EAC had the crucial role as the Client of the project, and their input has been invaluable. The various National Task Forces and the Regional Task Force was the real doers in implementing the processes at various level. NORAD, through their Embassy in Kampala, played a crucial role as funding agency, the World Bank as a coordinating entity and SIDA for constructive contribution during the processes and additional funding to the National Task Forces.

References

Abila, R.O. and Jansen, E.G., 1997. From Local to Global Markets: The Fish Exporting and Fishmeal Industries of Lake Victoria - Structure, Strategies and Socio-economic Impacts in Kenya. Report No. 2, IUCN The World Conservation Union, Eastern Africa Programme, Nairobi, Kenya.

Balirwa, J.S., Chapman, C.A., Chapmann, L.J., Cowx, I.G., Geheb, K., Kaufman, L., Lowe-McConnell, R.H., Seehausen, O., Wanink, J.H., Welcomme, R.L. and Witte, F., 2003. Biodiversity and Fishery Sustainability in the Lake Victoria Basin: An Unexpected Marriage. *Bioscience,* **53(8):** 703-715.

Bootsma, H.A. and Hecky, R.E., 1993. Conservation of the African Great Lakes: A limnological perspective. *Conservation Biology,* **7:** 644-656.

Bulloch, J. and Darwish, A., 1993. Water Wars: Coming Conflicts in the Middle East. St. Edmundsbury Press, London, pp. 124-141.

Costanza, R., d'Arge, R., de-Groot, R., Farber, S., Grasso, M., Hannon, B., Limburg, K., Naeem, S., O'Neil, R., Paruelo, J., Raskin, R., Sutton, P. and van den Belt, J., 1997. The value of the world's ecosystem services and natural capital. *Ecol. Econom,* **25(1):** 3–15.

COWI, 2002. Integrated Water Quality/Limnology Study for Lake Victoria. Lake Victoria Environmental Management Project Report, Consulting Engineers and Planners.

Goldschmidt, T., Witte, F. and Wanink, J., 1993. Cascading effects of the introduced Nile perch on the detritivorous/phytoplanktivorous species in the sublittoral areas of Lake Victoria. *Conservation Biology,* **7:** 686-700.

Goldschmidt, T., 1996. Darwin's Dreampond: Drama in Lake Victoria. MIT Press, Cambridge, MA, USA.

Johnson, T.C., Scholz, C.A., Talbot, M.R., Kelts, K., Ricketts, R.D., Ngobi, G, Beuning, K., Ssemmanda, I. and McGill, J.W., 1996. Late pleistocene desiccation of Lake Victoria and rapid evolution of Cichlid fishes. *Science,* **273:** 1091-1093.

Kaufman, L., 1992. Catastrophic change in species-rich freshwater ecosystems: The lessons of Lake Victoria. *BioScience,* **42:** 846-858.

NBI, 2001. Strategic Action Program: Overview. Entebbe, Nile Basin Secretariat in cooperation with the World Bank.

Nicol, A., 2003. The Nile: Moving Beyond Cooperation. Water Policy Programme, ODI. SC-2003/WS/61.

Ntiba, M.J., Kudoja, W.M. and Mukasa, C.T., 2001. Management issues in the Lake Victoria watershed. *Lakes and Reservoirs: Research and Management,* **6:** 211-216.

Raphaeli, N., 2004. Rising Tension over the Nile River Basin. *MEMRI, Enquiry and Analysis*. 29 February, 2004.

Sanchez, P.A., 2002. Soil fertility and hunger in Africa. *Science,* **295:** 2019-2020.

Shiva, V., 2002. Water Wars: Privatization, Pollution, and Profit. South End Press, Cambridge, Mass.

Statkraft Grøner, NIBR, Cornell University, MISR, UCLAS and Agrechs, 2003. The Vision and Strategy Framework for Management and Development of Lake Victoria Basin. Main Report (Vol 1) and National Reports for Kenya, Uganda and Tanzania (Vols. 2-4).

Verschuren, D., Johnson, T.C., Kling, H.J., Edgington, D.N., Leavitt, P.R., Brown, E.T., Talbot, M.R. and Hecky, R.E., 2002. History and timing of human impact on Lake Victoria, East Africa. *Proceedings of the Royal Society of London,* **B 269:** 289-294.

6

Management of Headwaters in Acidified Areas along the West Coast of Norway

Atle Hindar, Yvan Orsolini[1] and Brit Lisa Skjelkvåle[2]
Norwegian Institute for Water Research (NIVA), Televeien 3
N-4979 Grimstad, Norway
[1]Norwegian Institute for Air Research (NILU), P.O. Box 100
N-2027 Kjeller, Norway
[2]Norwegian Institute for Water Research (NIVA), P.O. Box
173 Kjelsås, N-0411 Oslo, Norway

Introduction

Management of headwaters aims at achieving good ecological status, and global, European and local options exist for improving the quality of acidified waters along the west coast of Norway. Norwegian headwaters are often considered clean as they are situated in areas far from or up-stream of large pollution sources. However, long-range transported air pollution has impacted surface waters in southern Norway for decades, and the resulting acidification and die-back of thousands of fish populations are widely known.

Management of acidified waters is a complex challenge and relatively straight-forward at the same time, depending on the management strategy. The aim of reducing acid deposition is mainly handled through international negotiations on emission reductions, as reflected in the Gothenburg protocol of 1999. Local options exist if the aim is to improve the acidified waters at the site.

Lakes and rivers along the west coast of Norway are acidified, most severe in southern parts and increasingly less obvious going north. This area is characterised by thin soils and a fjord-like topography, and receives heavy amounts of rain, typically between 2.5 and 4 m per year. Extreme weather, as characterised by storms from the west, transports large amounts of sea-salts in-land. Ion-exchange of part of the sodium from these salts in the soils

(Wiklander, 1975; Skartveit et al., 1980), as revealed by negative calculated non-marine Na concentrations in surface waters, may have dramatic effects. In acidified areas, inorganic aluminium and H^+ ions are transported to the streams and rivers, and regional fish kills have been reported (Hindar et al., 1994).

Extreme weather in western Norway is related to climatic patterns over the Euro-Atlantic area. The most important and widely known of these patterns is the North Atlantic Oscillation (Hurrel, 1995; Marshall et al., 2001), but other patterns may also be significant for the weather situation. As more extreme weather is expected in the future as a result of global warming, we have coupled long-term river-data with climatic patterns to reveal dominating and interacting trends for water quality in rivers and catchments. Of special interest is the interaction of heavy sea-salt deposition, related to extreme weather-conditions, with acidification status and trends.

To avoid potentially toxic conditions in salmon rivers during extreme weather events, terrestrial liming experiments have been conducted in the area. If successful, such locally based measures might represent an option to achieve good ecological status, as requested by the EU Water Framework Directive.

This work is mainly based on three already published papers, and we refer to them for more details regarding monitoring data, trends, sea-salt effects, relation to climatic patterns, and effects of terrestrial liming. The situation of sites included is shown in Figure 1.

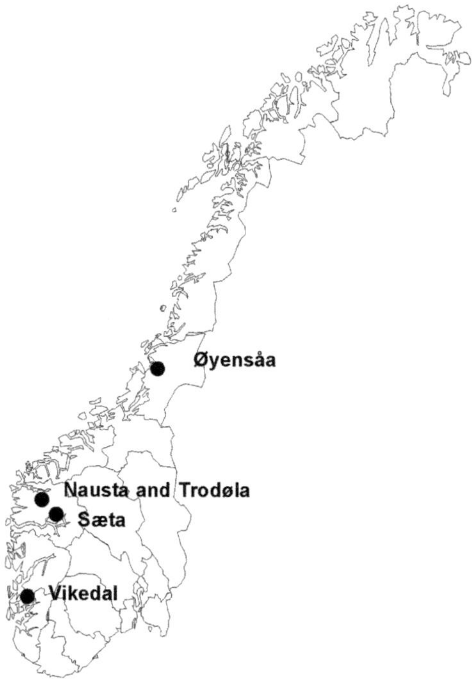

Figure 1: Situation of monitored rivers.

Threats and Trends

Long-term data from the monitoring of long-range transported air pollutants and effects on lakes, rivers, streams are mainly from the early 1980s. Trends in acid deposition are relatively uniform over southern Norway (Figure 2), and a more than 50% reduction in lake sulphate from 1986 until 2003 has been found. Nitrogen-deposition also decreases, but very slowly over the same period.

Trodøla is a tributary to River Nausta in the Sogn and Fjordane County in the northern part of the west-coast area. In this pristine stream pH has increased

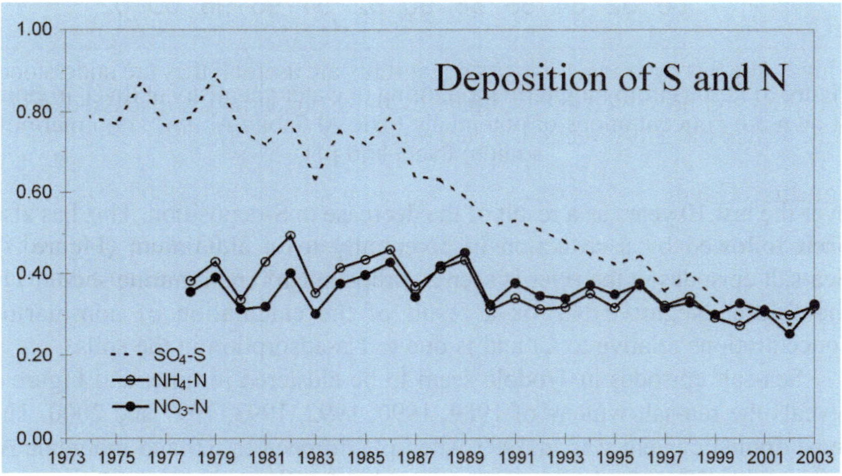

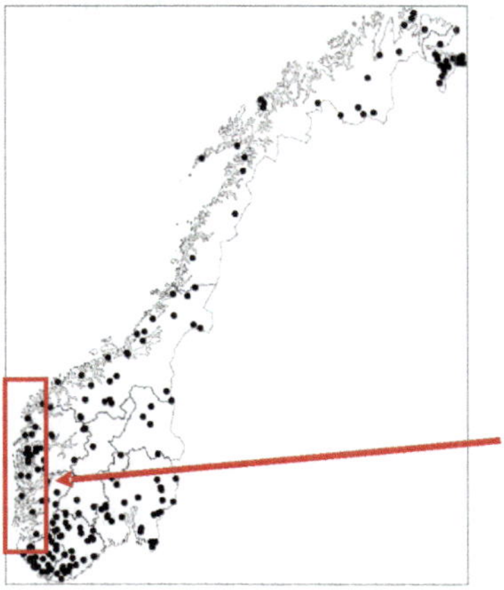

		% reduction in sulphate 1986-2003
I	East; North	49
II	East; South	60
III	Mountains; South	57
IV	South; East	56
V	South; West	54
VI	West - South	56
VII	West - North	48
VIII	Middel	40
IX	North	45
X	Finnmark; East	30

Figure 2: Yearly volume-weighted means of sulphate, nitrate and ammonium in deposition for six representative monitoring stations in south+east (NILU, 2004; upper panel) and the reduction in lake-SO_4 from 1986 to 2003 in different regions in Norway (lower panel).

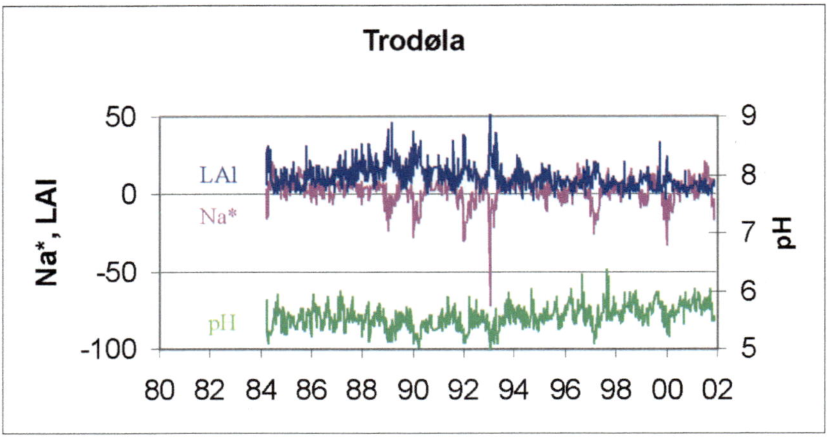

Figure 3: Results from long-term monitoring of water chemistry at River Trodøla. Shown are concentrations of potentially toxic Al (labile Al; LAl), non-marine sodium (Na*) and pH.

over the last 10 years as a result of the decrease in S-deposition. This has also been followed by a reduction in potentially toxic aluminium (Figure 3). Sea-salt episodes in the river is seen as drops in Na* (non-marine sodium) in the figure. Negative Na* is a result of the calculation of non-marine concentrations relative to Cl and is due to Na-adsorption in the soils.

Sea-salt episodes in Trodøla seem to be clustered in time, and Figure 3 reveals the sea-salt-winters of 1989, 1990, 1992, 1993, 1997 and 2000. The most dramatic event was in 1993 when pH dropped to 5.0 and inorganic Al increased to 50 μg/L. At this time regional fish kills in rivers along the coast was documented (Hindar et al., 1994).

Climatic Patterns

Extreme weather is driven by climatic patterns over the Euro-Atlantic area. Hindar et al. (2004) found that sea-salt episodes were strongly related to the positive phases of the North Atlantic Oscillation (NAO; Figure 4). In the Trodøla and Sæta rivers in Sogn and Fjordane County (Figure 1), Cl followed the same pattern as the winter NAO index in the period 1985-2000 (Figure 5). The 75-percentiles seem to give better fit than the mean or 99-percentile due to low or high, respectively, representation of very high and extreme numbers.

In this work a direct link between the NAOI and potentially toxic concentrations of aluminium in rivers was documented. Measured inorganic Al in spot samples was grouped in classes with increasing concentrations and the corresponding monthly NAOI for these occasions were averaged. Significantly different means of NAOI (Table 1) showed that potentially toxic water occurred at times when the antecedent, monthly NAOI was high. Inorganic Al (LAl) is categorized in concentration classes, and significantly

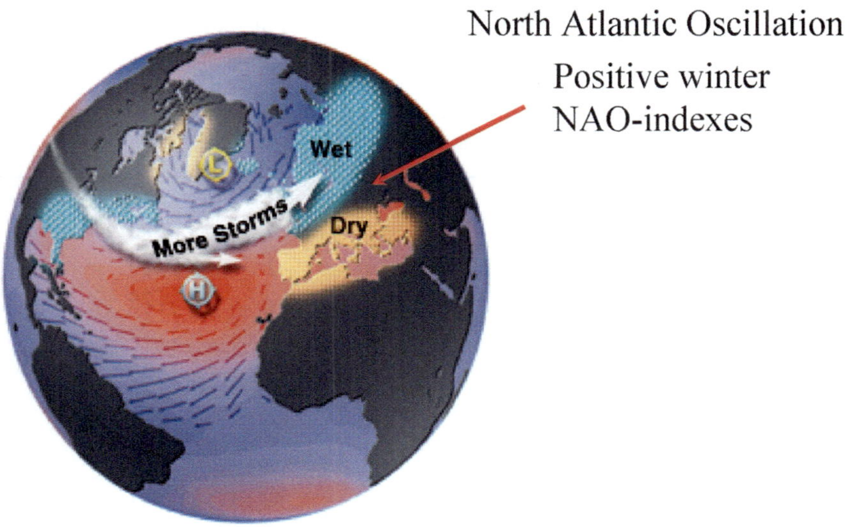

Figure 4: The North Atlantic Oscillation in the positive phase associated with extreme weather along the west coast of Norway. (Source: http:www.cru.uea.ac.uk/~timo/projpages/nao_update.htm)

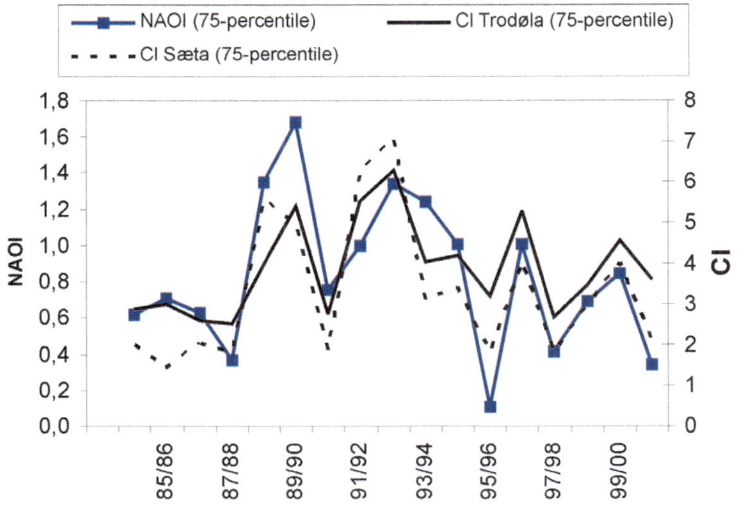

Figure 5: The NAO index and sea salt concentrations, represented by chloride, in the Trodøla and Sæta rivers.

different means (sdm) of monthly NAOI are identified by ANOVA and marked with different small letters. Potentially toxic Al-concentrations for Atlantic salmon smolts are in italics. Numbers in each class is given.

Recent work has shown that not only the NAO, but also other climatic patterns of the Euro-Atlantic sector has a significant impact on water quality and explains even more of the variation in Cl than the NAOI alone (Hindar and Orsolini, unpublished data).

Table 1: NAO indexes corresponding to inorganic Al in spot samples from four rivers along the west-coast of Norway

River	*LAI-cat.* $\mu g\ L^{-1}$	*NAOI-M* *mean*	*sdm*	*n*
Vikedal	0-1	-0.29	a	28
	11-20	0.27	b	24
	21-40	0.10	b	39
	>40	0.83	c	25
Nausta	0-8	-0.06	a	94
	9-15	0.13	a	46
	>15	0.69	b	26
Trodøla	0-10	-0.04	a	201
	11-20	0.21	b	164
	21-30	0.66	c	53
	>30	0.65	c	17
Sæta	0-10	-022	a	46
	11-20	0.05	b	50
	>20	0.39	c	41

Management Options for Good Ecological Status

Water quality dynamics in the headwaters along the west coast are governed by a combination of acidification impact and climatic variability. To modify these stressors one has to rely on global and European negotiations on emission reductions of S + N oxides (acidification) and CO_2 + other gases that may influence the climatic patterns. Results from the international co-operation are currently reflected in the Gothenburg protocol of 1999 and the Kyoto protocol ratified in 2005. Further emission reductions, although not so large as seen over the last decade, is expected for S and N, but the situation for CO_2 seems to be less optimistic.

Local measures are mainly different liming strategies for source areas and surface waters (lakes and rivers). Experiments with terrestrial liming have been conducted in the area in order to examine if sea-salt derived Al-mobilisation may be reduced. Addition of 760 metric tons of coarse-grained dolomitic limestone in the Brommeland catchment (3.9 km^2) in the Suldal River system (Figure 6) in 1999 had a rapid and adequate effect over time on stream-water chemistry (Hindar, 2005). pH increased to above critical levels of 6.0 and inorganic Al was reduced to below 20 μg/L, regarded as a critical value for salmon smolts.

The change in water chemistry may be exemplified by looking at the sea-salt-winters of 1997 (before liming) and 2000 (after liming). The episodes during these winters were comparable when it comes to Cl-increase, Na*-reduction and duration, but completely different for stream inorganic

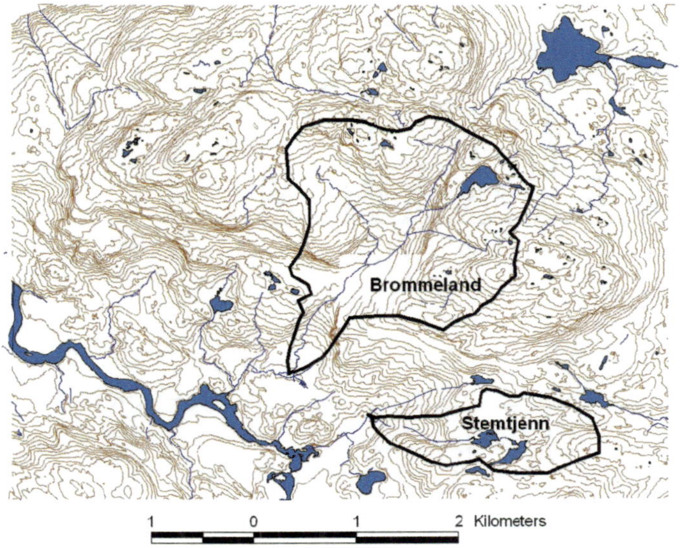

Figure 6: The limed catchment (Brommeland) and reference catchment (Stemtjenn) in the lower part of the Suldal watercourse. Water flow was measured in the stream north of Stemtjenn.

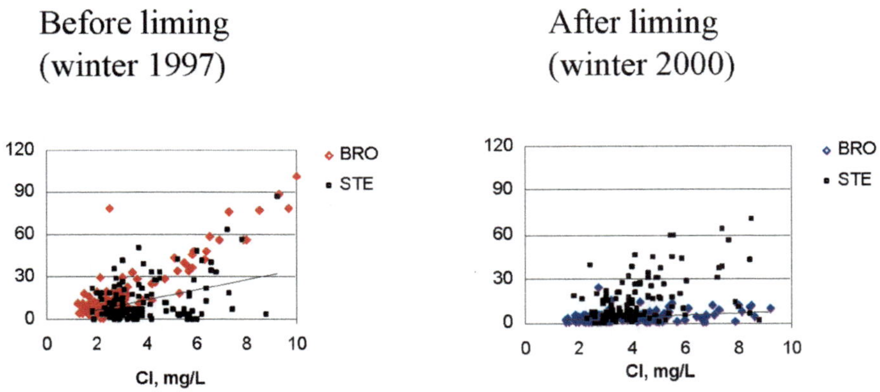

Figure 7: Chloride drives toxic aluminium, but not after liming; Relations between chloride (Cl) and inorganic Al (labile Al; LAl) before (left panel) and after (right panel) whole-catchment liming.

Al concentrations in the limed catchment. Cl changed from being closely related to Al-increase before liming to having almost no impact after liming (Figure 7). Details from the winter 2000 (Figure 8) reveal rapidly changing runoff, steadily increasing Cl-concentrations and no Al-mobilisation after liming. Similar experiments and modelling of future trends have shown that adequate liming effects may last 40-50 years (Hindar et al., 2003).

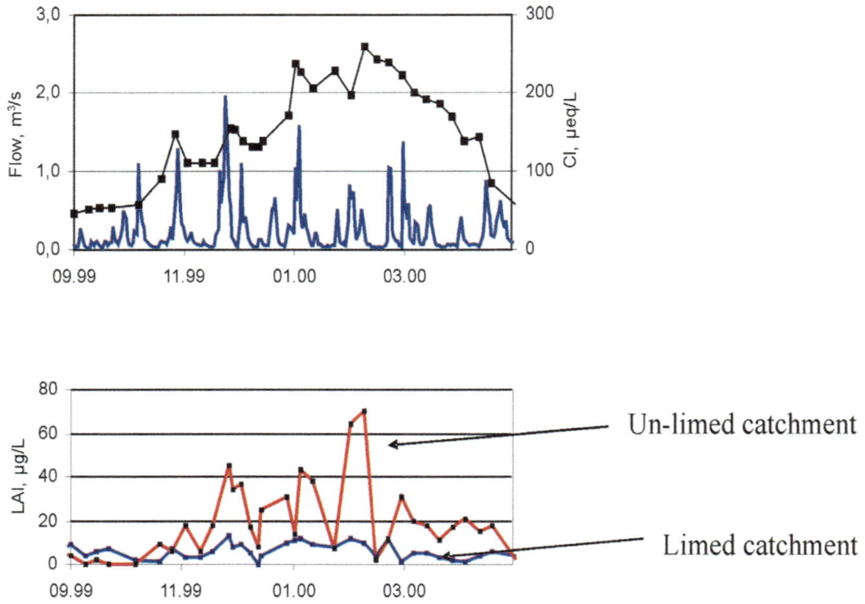

Figure 8: Flow and stream Cl (upper panel) and stream inorganic Al (lower panel) concentrations in the Suldal catchments during the winter of 1999-2000.

A Look into the Future

Future trends in water chemistry will be governed by a combination of further effects of emission reductions and on climatic variability. One important part of this will probably be a delayed effect of reduced acid inputs. Dynamic modelling has shown that soil properties and soil processes have major impacts on how fast recovery will take place after certain emission reductions. A confounding factor is also the effect of the long-term nitrogen deposition and an expected decrease in the C/N ratio of soils. Empirical data from different sites in Europe have shown that nitrogen leaching tend to be higher in areas of lower C/N ratio. Increased leaching of inorganic N may offset the effects of decreased S and N deposition.

We are currently exploring how climatic variability may interact with these future trends. Higher temperatures and reduced snow-cover may impact nitrogen processes in soils and thereby result in increased N leaching. But increased temperature may also enhance growth and thereby promote N retention. The net effect may be unclear and dependent of other factors again.

More severe weather in the future may be a mix of more frequent and more extreme storm events. Thus sea-salt deposition may be expected to happen more often and with higher stream Cl-concentrations. "More often" may result in a better memory in the soils of the last episode, which means that the pool of available toxic Al is lower. This again may imply that the sea-salt effects may be less severe. However, higher concentrations as such increase the potential of Na-adsorption and also of getting H^+ and Al-ions out in the streams and rivers.

An effect of acidification is leaching of Ca and Mg from the soils together with the strong acid ions. This long-term leaching is now seen as decreased levels of these ions in lakes and rivers. Terrestrial liming may thus represent not only a necessary measure for improving the water quality for sensitive aquatic biota, but also as a measure for supplying the soils with lost ions. A long-term management for these areas may therefore be a mix of emission reductions at the global and European scale and local measures such as terrestrial liming.

Acknowledgements

This work is based on monitoring data from the national monitoring programme for long-range transported air pollutants and its effects and results from two projects (Klima Salt and the Terrestrial Liming Project) under contracts from the Directorate for Nature Management (DN). We will especially acknowledge Else Løbersli and Roy Langåker at DN for their support.

References

Hindar, A., 2005. Whole-catchment application of dolomite to mitigate episodic acidification of streams induced by sea-salt deposition. *Sci. Tot. Environ.,* **16.**

Hindar, A., Henriksen, A., Tørseth, K. and Semb, A., 1994. Acid water and fish death. *Nature,* **372:** 327-328.

Hindar, A., Tørseth, K., Henriksen, A. and Orsolini, Y., 2004. The significance of the North Atlantic Oscillation (NAO) for sea-salt episodes and acidification-related effects in Norwegian rivers. *Env. Sci. Technol.,* **38:** 26-33.

Hindar, A., Wright, R.F., Nilsen, P., Larssen, T. and Høgberget, R., 2003. Effects on stream water chemistry and forest vitality after whole-catchment application of dolomite to a forest ecosystem in southern Norway. *Forest Ecol. Manage.,* **180:** 509-525.

Hurrell, J.W., 1995. Decadal trends in the North Atlantic Oscillation: Regional temperatures and precipitation. *Science,* **269:** 676-679.

Marshall, J., Kushnir, Y., Battisti, D., Chang, P., Czaja, A., Dickson, R., McCartney, M., Saravanan, R. and Visbeck, M., 2001. North Atlantic climate variability: Phenomena, impacts and mechanisms. *Internat. Jour. Climatol.,* **21:** 1863-1898.

Skartveit, A., Halsvik, B. and Meisingseth, E., 1980. Deposition of Sea-salts and Acid Components—Transport of Ions in Catchments in Western Norway (in Norwegian); SNSF-Project, IR 63/80, 82 pp.

Wiklander, L., 1975. *Geoderma,* **14:** 93-105.

Section B
Catchment and Streamflow Hydrology

7

Monitoring for Modelling Reality and Sound Economics

Ian Cordery and Peter S. Cloke[1]
School of Civil and Environmental Engineering
The University of New South Wales, Australia
[1]NSW Department of Commerce, Australia

Introduction

The economics of water resources monitoring has been examined several times in recent years (Cordery and Cloke, 1992; Peck and Teisberg, 1993; Green and Herschy, 1994; Moss, 1970; Stewart, 1994). The motivation for these examinations seem to have been to show that the cost of monitoring is small compared to the benefits (more than nine times the costs) that are being derived from the archive of data. However, though all the published studies show that the benefits of monitoring are in fact large compared to the costs of the monitoring and maintaining of the archived data, monitoring of water resources has steadily declined almost universally, and continues to decline (Cordery, 2003). It would appear that the only country where water resources monitoring has increased in recent years is New Zealand. The cause of the continued decline seems to be that data collection (monitoring) is seen as a public activity, and there is competing needs for public funds in other areas where the connection with public benefit is directly obvious and immediate. The community is not aware of the link between collection of hydrological data and the costs of public infrastructure. If the public was aware that the benefits of monitoring are as large as suggested in the studies cited above it would seem likely that there would be a higher demand to monitor phenomena that have large effects on infrastructure costs. This must mean that leadership is needed to communicate with and convince the direct users of data, then the public and ultimately those approving the public expenditure programmes. In the authors' opinion logical argument has seldom achieved quantum changes. Change only appears to be achieved in response to a significant event which causes damage, usually combined with leadership that can champion the solution.

How to Encourage Increased Monitoring?

Lack of realization of large demonstrated potential benefits is not unique to monitoring of water resources. Parallel situations abound in other areas of endeavour. For example we are tempted to think that the large benefits of "wonder" drugs and profits to multi national pharmaceutical companies result from altruistic research within those companies. Yet in most cases the discovery and initial testing of the drugs occurred in under-funded University laboratories where staff and facilities are in short supply and are funded by grants obtained only by hard endeavour and commitment of dedicated researchers in a very competitive environment. The pharmaceutical companies usually only become involved when they see a "winner". However most of the developments (and risks) occurred in a (poorly) publicly funded environment. Presumably many large potential public benefits never reach the public, not from lack of efficacy, but from lack of perceived benefits to company bottom lines.

The hydrological monitoring situation is in many ways parallel to the above example. However, while the public benefits of data collection are large, we are unable to conceive of means by which a company that collected hydrological data could derive profits from the data—the benefits are of a type that contribute to the whole community. A parallel could be the "quit" smoking campaign where benefits are community-wide health improvements but it is difficult for significant profits to be added to any company or even government bottom line. Therefore these activities of necessity must be publicly funded. However, arousing sufficient public interest to ensure the necessary funding is no easy task.

Educate the Public on the Vital Role of Data

Monitoring water resources is a little like collecting community statistics, say in surveys or in a census. Researchers can use the statistics to examine the magnitude of social changes, such as the change from single to double income families, or the causes of human illness, disability or death, or causes of social ills, which can lead to recommendations of changes in public policy to reduce, or in many cases prevent those undesirable outcomes. Water resources data provides a basis for understanding of the local water resources situation and for engineers and managers to design for safe provision of adequate, high quality water supplies, treatment and environmentally acceptable recycling of wastes, and provision of infrastructure to support these activities that will themselves not add to the risks to safety of the community, or adversely impact on the environment.

Demonstrate Water is a Scarce Resource

Water is essential for life and for most human activities, including making corporate profits. Until recent years water resources have been relatively abundant, compared to demands for access to them, in most parts of the world. However increased population, high expectations of living standards and huge increases in irrigated farming have changed the equation to the point where water is a scarce resource in many regions, and the regions where available supplies exceed demand are reducing daily. As a result considerable expertise is now required to manage the scare resource—to assess how much water is available and how much can be consumed without compromising environmental values. Expertise is also required to design storages and spillways, to define the extent of flood prone land, to treat and recycle wastes and other quality compromised water and to minimize the impacts of changes in land use on the water environment.

All the expertise available cannot reliably predict outcomes and safely and optimally manage these activities without factual information, i.e. observed recordings of rainfall, flows, groundwater levels and quality characteristics. Basic data are needed to allow assessment of the likely outcomes of proposed developments and changes, including proposals for improvement of the general state of (degraded) water resources and the environment.

Show that Data are Needed before Water Resources Development Proposals can be Made

There are many similarities between development of good management of water resources and improvements in management of human health. In human health management, laboratory studies of biochemistry and psychology can suggest avenues for further investigation, but the laboratory studies are themselves totally inadequate as panaceas for bringing about a desired improvement in health. The human system is so complex and so poorly understood that trials must be conducted and statistics collected from real people to attempt to determine the actual outcomes (as distinct from those predicted from laboratory studies) of any chemical or physical interventions. When predicted outcomes look promising, corporate bodies invest huge resources into the conduct of carefully controlled clinical trials. These are conducted with sufficiently large number of subjects to produce results, which can then be statistically analysed to provide convincing evidence of the efficacy of the proposed interventions.

In the world of water resources it is rarely possible to conduct "laboratory studies". Field observations can be conducted but "clinical trials" such as are used in health investigations are rarely possible. The only feasible means of obtaining hard evidence is to examine the information that has been collected over the years from a range of locations and hopefully, using similar statistical analysis techniques to those applied to results from clinical trials, to draw out the evidence to support the proposed management changes. Unfortunately in the water resources professions it is not possible to gather subjects for clinical trials (even though in the health field researchers and managers complain of the length of time and expense of such activities, the trials are possible) but one must somehow find occasions and locations in the landscape (speaking here very broadly) where observations of relevant physical and chemical variables have actually been collected in the past and the measurements archived, which can be used to (objectively?) assess the merits of the proposal. Measurements from the past are emphasised here because most events of water resources interest are triggered by the occurrence of liquid water, either in rainfall or snowmelt events. As we all know these events, being natural, do not occur to order, nor do they occur regularly. Hence our "clinical trial" must depend on random, naturally occurring events that are observed at most a few times per year, but in some cases only once in tens of years. Therefore to build up a sample of observations from which statistically meaningful (significant) inferences can be drawn requires that the observations be made over many years. If we are proposing to construct, next year, a dam to store water for a new irrigation scheme, there is now no opportunity to obtain any meaningful statistics on the likely streamflow into the reservoir. By the time construction is complete we may have at most three years of observations. But were the flows in those three years in which data could be collected average, below average, were there exceptional flood flows? How can one answer these vital questions? Only by continuing to observe streamflows for many years—at least 20 and in highly variable climatic regions such as those around 30° latitude, for 30 years at the very least. Examples of problems of short data records will be given later. When most development proposals are made there is usually some delay before implementation during which time some relevant water resources information can be measured. However data collected under these circumstances is usually far from adequate for design of multi million dollar works, especially when it is remembered that the consequences of failure of some of these works could be the deaths of large number of people.

Therefore just about the only water resources data that can provide the needed statistically significant evidence to support many of the day-to-day and long-term decisions water resources managers must make are data that were observed in the past and are already contained in a data archive at the time the development proposals are made. Obviously it is not possible to have an archive containing the data needed to provide solid statistical backing for every possible development proposal that is likely to arise. However, as has been shown in several studies, the data currently available in most parts of the world are inadequate to provide the information expected of them. Firstly they are not able to provide the investors (the general tax paying population) with assurance that water resources infrastructure are being developed and designed for efficient use of public resources. Secondly, most archives of data do not contain sufficiently large samples of data to provide a high confidence in safety of design (for dams as an example) or the reliability of a water supply on which they depend or complete degradation of part of their living environment. Thirdly, most data archives are not sufficiently comprehensive to encourage adventurous thinking about how we can develop/use our environment to provide greater benefits with less adverse impacts.

Development Demands Unreasonably Dictate Policy

In most situations the demands for developments which impinge on water resources and the environment are compelling but the ability to scientifically assess likely long-term impacts, with support from reliable, verifiable evidence is very limited because the measurements of landscape and environmental variables needed to provide this evidence have never been made at or near the site of interest. So, in the absence of strong contrary evidence the developments go ahead, and in many cases future generations have to face the consequences of these inadequately investigated decisions. The precautionary principle needs to be applied much more widely. However the pressures for more water for agriculture in particular are so great that the precautionary principle is given lip-service at best. If more water resources data were collected and archived there would be less need for application of this principle, as the environment and the water component of it could be better understood and more sound bases would be available for assessing water resources development proposals.

Examples of Problems from Sparsity of Data

There are endless examples, some small and local, some of national and international significance. Many international water disputes, while being dominated by politics, are usually underlain by either a lack of factual hydrological information or by secrecy about, or alleged clandestine manipulation of data in the border regions.

In Australia, an (large) island nation with no land borders, there has been innumerable inadequate decisions which reflect the problems and costs of inadequate hydrological data. We will give following examples.

Western Australian Salinisation

Practically no systematic hydrological flow data (surface water or groundwater) were collected in the west of Australia prior to 1950. It was not realized that in its natural, pre-development state, the region that is now called the "wheat belt", which has mean annual precipitation in the range 400-700 mm and covers about 100,000 km^2, had an actual annual excess of precipitation over evapotranspiration of less than 10 mm. It was not even known, for example, that what is now called the Avon River system, which flows into the central water feature of the city of Perth, had any outlet to the sea. From the late 1940s large number of otherwise unemployed families were encouraged (given money) to settle in the area, clear it of its native vegetation and produce wool and wheat, for which, at the time there was huge worldwide demand. Clearing of forest and the growing of grass or cereals changed the hydrological balance and resulted in rapid rise of the water table across the region. Within 20 years salty water had risen to the surface over thousands of hectares and the productivity of much of the land began to decline. Lack of understanding of the hydrology (in large part due to there being no hydrological monitoring) and of the salt storage in the soils of the region has caused widespread devastation. Salt has accumulated in the soils over geological time. At the west coast the rain has an average 9 mg/L of sea salt. With practically no flow of water from the region above or below ground, as the soil water evaporates most of the salt accumulates in the soils. The change in the hydrological regime produced by land clearing has led to extensive water logging and surface salinisation. Several of the small towns which service the farming communities have groundwater just below the soil surface (this is a very flat region) so that all below ground amenities are water logged and subject to salt attack and all buildings with basements have them full of water and all masonry and concrete is in a continual state of decay. All these economically and socially disastrous effects were the result of misguided assumptions about the land systems, which could not be contradicted (though some tried) because no systematic hydrological data collection had been instituted. Water for human and stock drinking was collected in a few surface storages in a small range of hills on the western edge of the region and distributed in small aboveground pipes which were installed as part of the government incentive scheme to encourage development. Even this infrastructure is now under threat from salt attack and needs ever more expensive maintenance. It is now virtually impossible to reverse the effects of the salinity, and the costs of doing so would be many times greater than all the benefits that have accrued from the cleared land in the last 60 years. Early collection of hydrological data could have forewarned of potential problems and lead to different, less salt vulnerable patterns of development.

Water Supply Security

During 2002-03 large parts of Eastern Australia suffered the worst drought in the last 100 years and as a consequence of that drought Sydney is still subject to restrictions on water use. Water supply security of many of the towns is being reviewed in the light of the drought and their performance during that drought. However such reviews will be constrained by the available hydrological data as illustrated by the following examples.

Example One

A town water supply scheme serving a population of some 13,000 on the Eastern coast of Australia came close to failing. Essentially the scheme relied on pumping from an alluvial aquifer that was directly fed from an adjacent river. Observations during the drought suggested that the aquifer had smaller storage than previously judged from examining limited data and that the aquifer was leaking. As a consequence of the drought and growing demand, options to provide appropriate drought security are being investigated. The main option involves providing an off-stream storage to be supplied from pumping surplus water from the river (after environmental objectives have been satisfied) and to use this during dry times when there is insufficient water in the river to meet demand.

Rainfall records dating back to 1890 are available for the area. A gauging station has been in operation on the river since 1959 downstream of the off-take. The catchment to the off-take comprises 60% of the catchment to the gauging station. While the streamflows are longer than is often the case they have been affected by upstream abstractions for water supply and irrigation. Records of town water supply were only available back to 1993 and no data were available on irrigation abstractions. Hence, even with over 40 years of streamflow observations estimation of flows under natural conditions required much imagination. Using the streamflow data, a rainfall runoff model was calibrated and the streamflow data extended using the longer rainfall records. The extended streamflow data was then used in a model to determine the storage size needed to meet future demand with appropriate drought security while satisfying environmental flow objectives.

Use of the streamflow sequence based on the model calibrated with 45 years of recorded streamflow resulted in a storage size of some 5400 ML while use of a streamflow sequence based on the model calibrated with some 10 years of naturalised recorded streamflows resulted in a storage size of 6200 ML (for a given set of environmental flow objectives). This difference in storage size equates to some $2 million dollars, which is about 20% of the prime cost of the storage. This amount is quite significant for the small community that would have to pay for the storage to provide adequate drought security for their water supply.

A side issue on the above example is that it is not just a matter of collecting the traditional hydrological data but also of recording data on parameters that impact the hydrological data such as abstractions (e.g. water supply and irrigation) and land use changes.

Example Two

A small village of some 300 people in Eastern Australia that relied for its water supply on pumping from a pool created by a weir across a river into a small off-stream storage of some 25 ML was looking to upgrade its security of supply to meet future growth. The current demand was some 80 ML/a whereas its future demand was expected to be some 95 ML/a. Unfortunately no data had been collected at the weir. Using a standard rainfall runoff model with some 100 years of available rainfall records, and using standard procedures that have been developed to estimate the model parameter values for the ungauged catchment, some one hundred years of historic streamflows were synthesised. Use of these streamflows in another model to assess the water supply security indicated that the current system could be expected to meet the future demand on a secure yield basis.

However due to the uncertainty in estimating model parameter values for ungauged catchments an alternative streamflow sequence was synthesised utilising limited flow data at a gauging station in the same river basin but with a much larger catchment and then transposing those data to the weir site using judgement for differences in catchment rainfall and area. Use of these streamflows in the model to assess the water supply security indicated that the current system was vulnerable to droughts that had occurred in the past and the off-stream storage capacity would need to be substantially increased (by up to 80%) to meet the future demand on a *secure yield* basis.

Without any appropriate data, using two acceptable procedures result in two very different conclusions. In this case the decision has been taken to accept the current situation with the possibility of more frequent restrictions in the short term and to collect some data (for two years) at the weir to allow limited model calibration towards an assessment for reduced uncertainty. It is ironic that the costs of the studies to date, which will probably have to be repeated after collection of two years of data, probably exceed the costs that would have been involved in recording data at the weir for 10 or more years.

Flood Security

Dams in Australia generally have to be designed to be able to safely pass the probable maximum flood (PMF) if failure of the dam would result in a large population at risk (>1000 persons) for downstream residents. Smaller design floods are being applied to dams with smaller population at risk. Many dams are having to be upgraded resulting in total costs of hundreds of millions of dollars as a consequence of estimates of design floods being increased due to advances in the science of estimating the probable maximum precipitation (PMP) from larger data sets of recorded extreme storms and supposed advances in the methods of transposing the PMP to PMF. The supposed advances are based on more data. However since the actual data is still limited, the claimed improvements in estimates cannot be proven to be real. Generally the costs of upgrading a dam to accommodate an increased design flood are substantially more than if the dam had been originally designed to accommodate the increased flood. Due to the high costs of upgrading the numerous dams in Australia, current practice is to identify a design flood range generally from the 1 in 1000 year flood to the PMF (generally guessed to be about the one in a million year flood) depending on the population at risk if the dam failed from flood.

In many cases millions of dollars are being spent based on very limited data, since many of the dam catchments are ungauged. The following two examples illustrate the issues arising from the limited data.

Example One

A dam in Eastern Australia with a catchment area of some 100 km^2 was required to be upgraded to accommodate the PMF. Using the standard best approach of calibrating a standard rainfall runoff model using observed flood records the PMF was determined to be some 7000 m^3/s. Note that in this case flood data had been recorded for the dam catchment. If no data had been recorded for the catchment, using the standard procedures for estimating the PMF for ungauged catchments, which involves using regional equations for estimating the model parameter values, would have resulted in a PMF estimate of less than half (i.e. some 3000 m^3/s). According to the standard procedures for estimating extreme flood frequency curves (which of necessity are not based on recorded data), this flood was estimated to be 5000 times more likely to occur than the PMF. Even allowing for the uncertainty in the regional equations one could have expected a PMF estimate between about 2500 m^3/s to 4500 m^3/s, but real data suggested it to be of the order of 7000 m^3/s.

Example Two

A dam in Eastern Australia with a catchment area of some 150 km^2 was required by the dam safety regulators to be upgraded to meet current flood standards. No observed flood data were available for the dam catchment. Using a selection of standard acceptable practices for estimating floods for a dam catchment where no data are available resulted in the dam being assessed as being able to safely pass a flood with annual exceedance probability ranging from somewhere between about 1 in 400 to 1 in 15,000. This range is a consequence of no data being available, but it introduces considerable uncertainty as to the urgency of upgrading the dam or perhaps even introduces doubt about whether it really needs to be upgraded, depending on the potential for loss of life.

Conclusion

Far too little hydrological monitoring is occurring in the world today. A legitimate reason is not cost. Hydrological monitoring has repeatedly been shown to provide economic benefits which are far greater than the costs of observing, editing and archiving the resulting data. The reason has more to do with lack of understanding by water managers/allocators and the government officials and politicians to whom they answer, of the nature of water resources/hydrological processes. Monitored data make a vital contribution to any hydrological study. The less data are available the greater the uncertainty that must underlie decisions. Uncertainty in decision making in water resources systems can lead to either higher cost (to the community) or greater risk of failure of water infrastructure, as has been shown in examples given in this paper. The increased infrastructure costs or the cost of failures can be many times higher than the cost of any current monitoring programme or any that has existed. For public safety and good stewardship of public funds there should be more investment in hydrological monitoring.

Disclaimer

Statements and opinions expressed in this paper are those of the authors and do not necessarily reflect the views of their employer organizations.

References

Cordery, I., 2003. A case for increased collection of water resources data. *Aust. Jour. Water Resources,* **6:** 95-103.

Cordery, I. and Cloke, P.S., 1992. An overview of the value of collecting streamflow data. *Civ. Eng. Trans., Inst. Engrs. Aust. CE,* **34:** 271-276.

Green, C.H. and Herschy, R.W., 1994. Assessing the benefits of streamflow gauging. Conference on Economic Benefits of Meteorological and Hydrological Services. World Met. Org., Geneva. TD 630, 141-144.

Moss, M.E., 1970. Optimum operating procedure for a river gaging station established to provide data for design of a water supply project. *Water Resour. Res.,* **6:** 1051-1061.

Peck, S.C. and Teisberg, T.J., 1993. Global warming uncertainties and the value of information: An analysis using CETA. *Resource and Energy Economics,* **15:** 71-97.

Stewart, B.J., 1994. The value of hydrological data. Conference on Economic Benefits of Meteorological and Hydrological Services. World Met. Org., TD 630, 63-66.

8

The Nile Headwaters: Wetlands and Catchments in Highland Ethiopia

Adrian Wood, Ato Afework Hailu[1] and Alan Dixon[2]
Centre for Wetlands, Environment and Livelihoods, Huddersfield University, Queensgate, Huddersfield, West Yorkshire, UK
[1]Ethio Wetlands and Natural Resources Association
P.O.Box 1518, Addis Abebe 1110, Ethiopia
[2]Institute of Science & the Environment, University of Worcester
Worcester WR2 6AJ, UK

Introduction

Headwater wetlands are critical elements in the hydrological system. They help moderate stream flow, recharge groundwater resources and create micro-climates (Table 1). These functions are particularly important in tropical areas with seasonal rainfall as they help extend the availability of water.

The degradation and loss of headwater wetlands usually lead to increased flooding—often flash flooding, reduced dry season flows, lower groundwater tables and less availability of water and moisture during the dry season. Overall the loss of wetlands tends to exacerbate seasonal water shortages, and may have further impacts leading to desertification and biodiversity loss. These impacts will occur in the headwater areas but will also be felt to varying degrees downstream.

Table 1: Potential environmental functions and products of headwater wetlands

Hydrological	*Geomorphological*	*Biological*	*Climatological*
Groundwater discharge	Control of downcutting	Biodiversity maintenance	Micro-climate stabilisation
Groundwater recharge	Sediment retention	Wetland forest	
Water storage		Wildlife	
Flood control		Fisheries	
Flow moderation		Forage	

The Loss of Headwater Wetlands in the Nile Basin – A Neglected Issue

The Nile Basin Initiative (NBI) is seeking to improve the management of the Nile waters in order to improve the development benefits which can be gained from the efficient use of this river and its resources. The NBI recognises the importance of flow stabilisation, the need for reduced erosion and sedimentation in the basin and the need for groundwater recharge.

However, the NBI tends to focus on the role of dams for water storage and management, and sees soil and water conservation within the upper catchments as the main intervention in the headwaters. Wetlands in general are given limited attention except for the Sudd in the Sudan, and the role of headwater wetlands is not well addressed. This is despite the fact that, with extensive plateau areas in the Nile's headwaters above 1000 m amsl, somewhere in the order of 5% of this area is wetland, of one sort or another.

Figure 1: A sedge wetland in Ethiopia, typical of many parts of the Nile headwaters.

The loss of these headwater wetlands will make both the achievement of water flow moderation and sediment trapping more difficult and lead to increased desertification in these catchments. Due attention needs to be given to the processes of change which are going on in these wetlands as these areas can play a critical role in both the improved hydrological management and catchment rehabilitation sought by the NBI.

Any consideration of changes in the management of headwater wetlands must also recognise the important role which these areas play in the livelihoods of the people. There is a wide range of socio-economic benefits which are obtained from these areas when they are used with different degrees of transformation in these headwaters (Table 2).

Table 2: Socio-economic benefits of wetlands in various states of transformation

Pristine	*Part transformed*	*Fully transformed*
Spring water	Fodder	Dry season crops
Thatching	Thatching	
Medicinal plants	Water for stock	
Craft materials	Dry season crops	

Wetlands in Highland Ethiopia

Ethiopia provides over 80% of the water in the River Nile north of Khartoum from a catchment area of less than 20% of the total Nile Basin. The two main tributaries from Ethiopia are the Abbai (Blue Nile) and the Baro (Sobat). The former rises in the northern highlands, starting from Lake Tana, while the Baro drains the south-west highlands focusing on Illubabor, Masha and Bench Maji zones.

Within these catchments wetlands account for between 2% and 4% of the area, with the Lake Tana shore and associated swamps being the most extensive wetland area. There are other major wetlands in the Ethiopian highland tributaries, such as in the upper Didessa and in the Dabus, while there are extensive lowland wetlands in the Gambella area. Within all these headwater catchments there are also a large number of small wetlands, ranging from sedge swamps to seasonally flooded grasslands, many of which are in various states of degradation.

Ethiopia has seen the loss of many wetlands over the last few thousand years as a result of geomorphological processes, especially down-cutting into the plateau areas. This has been accelerated by devegetation, cultivation practices and heavy grazing which have all led to increased run-off and erosion. Where wetlands remain, a significant number have been degraded through grazing, cultivation or sedimentation. The outlines of past wetlands are clearly visible when flying over the country, while the loss of many wetlands within the last 50 years has been recently reported.

Specific Areas with Different Experience of Wetland Degradation

(a) South-West Forest Fringe – Illubabor Zone

Loss of wetlands is accelerating in this area as a result of population growth, the expansion of cultivation and government policies which encourage wetland drainage to improve food security. In some cases intensive cattle grazing has also had an impact and many sedge swamps are now just rough grazing land. In this condition they store little water due to soil compaction. In some locations in this zone there are apparently pristine, or non-transformed, wetlands which, together with the forest, provide important stabilising influences on the hydrology and micro climates. However, the overall trend is towards wetland degradation and loss as forests are cleared and wetlands drained for cultivation or grazing.

(b) Recently Degraded Western Highlands – Western Wellega Zone

This relatively recently deforested area, between the forests of the south-west and the long-deforested northern highlands, is subject to particular wetland pressures because of serious catchment degradation. The soils from the granitic basement rocks have been subject to severe erosion with the result that after less than 100 years of use the upland interfluves are severely eroded and virtually abandoned from cultivation. Now the main areas for cultivation are the valley bottom wetlands. However, food shortages are so severe in this area that these wetlands are completely cultivated every year – and used much more intensively than in Illubabor Zone to the south. There are serious concerns about how long these wetlands can sustain intensive drainage and cultivation, while the deposit of sediment from upland erosion is also reducing their productivity. The potential for a collapse in food production which would undermine food security is increasing all the time.

(c) Long Degraded Northern Highlands – Lake Tana Basin, Amhara Region

In the northern highlands the long history of settlement has seen progressive deforestation and land degradation over centuries. Wetlands have existed in

Figure 2: An intensively cultivated wetland in Western Wellega.

many of the tributary valleys of the Abbai and on the undulating plateau areas. However, downcutting by streams has led to damage of wetland sites, while cultivation has extended downslope in the plateau areas into the wetlands. Cultivation continues on the slopes and in the catchments despite the degraded nature of these highlands, and wetland cultivation has been quite limited, unlike in Western Wellega. However, grazing pressures have damaged these wetlands quite badly, while in recent years a new threat has arisen through the planting of eucalyptus trees in them to meet fuel and building wood needs. Many wetlands, including those along the shores of Lake Tana, have also been affected by sedimentation caused by the severe upland erosion, while weed infestation in wetlands is a new problem in these areas.

Processes and Dynamics of Wetland Loss in Ethiopia and the Nile Basin

These experiences with wetland loss are not unique to Ethiopia and are found to different degrees in other parts of the Nile headwaters, particularly Uganda, Kenya and Rwanda. Some of the key processes which need to be addressed in order to sustain the Nile's headwater wetlands include:

- Expansion of cultivation into wetlands,
- Intensive year-round grazing of wetland,
- Use of wetlands for fuel and timber lots,
- Catchment degradation, accelerated erosion and rapid down-cutting in wetlands,
- Sedimentation in wetlands from upslope erosion,
- New market opportunities which lead to increased pressure to use wetlands,
- Government policies, e.g. food security and resettlement which lead to use of wetlands, and
- Agricultural research whose results, such as short-maturing varieties of maize, may facilitate wetland use.

Sustainable Wetland Use and Rehabilitation – Balancing Needs

The rehabilitation of all wetlands within the Ethiopian Highlands to a pristine state, in order to improve the hydrological functioning of the Nile Basin, is not possible. There is a wide range of benefits which communities gain from the transformed or partially transformed wetland sites. These include: dry season cropping, seasonal grazing, fuelwood and building timber (Table 2).

However, the sustainability of some of these benefits is often in doubt due to soil degradation processes which are occurring, especially due to cultivation. Other negative side effects from some uses of wetlands include the drying up of springs, the removal of medicinal and thatching plants, and the loss of craft materials used by the poor, and a general desertification of the wetland environment and its surrounding catchments.

Hence, there is a need to try to develop some win-win situations by working with communities to:

(a) develop more sustainable wetland management practices, and
(b) undertake partial rehabilitation of wetlands to meet community needs.

These will involve identifying key driving forces influencing different wetland use scenarios. Initiatives in this area are already on-going in Ethiopia through the work of the Ethio Wetlands and Natural Resources Association in collaboration with Wetland Action and the Centre for Wetland, Environment and Livelihoods at Huddersfield University, UK.

Specific Regional Measures

South-west Forest Fringe

Work in this area focuses on the development of sustainable wetland management plans which try to maintain the agricultural potential of the wetlands whilst also ensuring that their role in hydrological moderation is maintained. Recommendations developed through discussions which draw on local knowledge and scientific research focus on:

- Maintaining portions of the wetland as natural swamp, to retain water and prevent downcutting;
- Minimising drainage channel depth and using drain blocking techniques;
- Single (rather than double) cropping, to minimise the drainage period;
- Blocking drains and outlets of wetlands to replicate annual flooding;
- Control of grazing, limiting cattle numbers as well as length and time of grazing;
- Integrated wetland and catchment management, with maintenance of forested catchments (often with coffee as an understorey); and
- Community capacity building to ensure the social sustainability of these practices.

Recently Degraded Western Highlands

The immediate need in this area is to reduce the pressure on wetland production so that these areas do not degrade further. This cannot be done without addressing the uplands where production must be revived. Measures proposed include:

- Stabilising the soil on the upper slopes and improve food production in these areas to reduce pressure for intensive wetland cultivation;
- Use of biological forms of soil and water conservation along with agro-forestry measures in these catchments;

- Development of composting and other organic farming methods to raise crop productivity and water infiltration and storage in the catchments;
- Strengthening of the capacity of existing indigenous organisations, such as the *Abba Lagas* – Wetland Management Committees, to address catchment issues as well as wetland management; and
- Use of wetland management techniques, similar to those in Illubabor Zone to develop more sustainable and hydrologically valuable wetland management.

Long Degraded Northern Highlands

Wetland rehabilitation is the main emphasis in this area. This involves working with communities to assess the land use needs in their areas and the changes in the use of wetland sites over recent decades. Where there are concerns that wetland rehabilitation can make contributions to the local community – such as for thatching material, dry season farming, fodder production, and well/spring recovery, plans are drawn up for rehabilitation of these areas. Specific measures use a combination of the catchment and wetland management methods developed in Illubabor and Western Wellega. One key is the need to always link to improved catchment and wetland management with soil and water conservation and land management measures which are economically attractive to farmers.

Key Elements of the Approach

A number of key elements in the approach developed by the three partners can be identified. These include:

Community Participation

All of this work starts from community stakeholder assessments of priority issues and needs, developing a clear picture of the local, community-based, knowledge about wetlands and catchments and their management, and making an assessment of the natural resource base and its use. The overall aim is, that by involving farmers and community structures from the outset, the capacity to address wetland and catchment management can be developed within the communities so that the need for external inputs is minimised and eventually removed over time.

Multiple Use of Wetlands

For wetlands to be maintained in some form they must be valued. Hence the different contributions which they make to the local economy and ecology have to be recognised. Overall it is most likely that wetlands will be maintained where multiple uses of these areas can be facilitated (Figure 3).

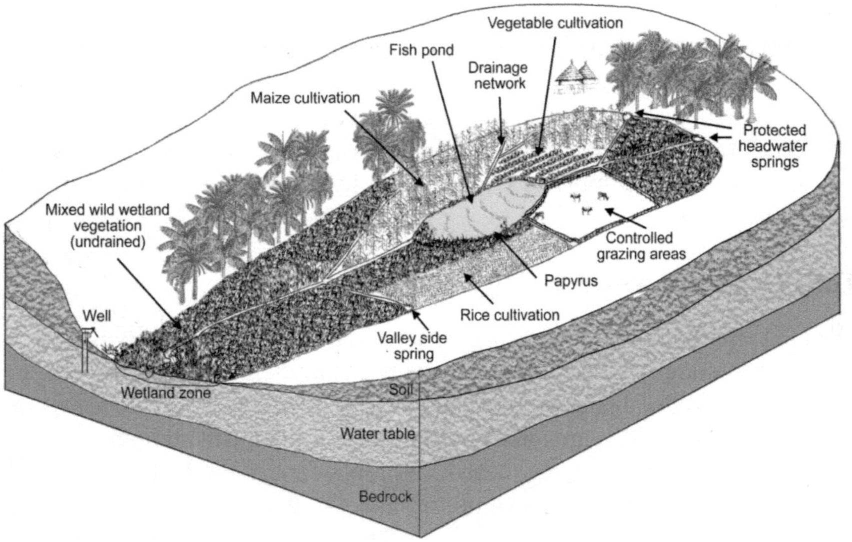

Figure 3: Multiple use model for sustainable wetland management in the Nile Headwaters.

Catchment and Wetland Integration

The importance of linkages between catchment and wetlands is well known. Successful wetland rehabilitation and management is dependent upon sound catchment management and hydrology. Hence this work always includes attention to upland land use and productivity, as well as the situation in the wetlands. Conversely, with improved hydrological functioning in the wetland and the lower catchment, wider catchment rehabilitation is facilitated.

Policy Dialogue

Given that government policies often impact upon wetlands, a major initiative of the on-going work in Ethiopia is to develop dialogues with the government at a variety of levels. This has already led to the establishment of wetland officers in Illubabor Zone and to a regional workshop in Amhara Region. In addition there have been national debates which have led to Ethiopia seeking to join the Ramsar Convention.

Strategic Planning

One future area for potential development is the establishment of a system wide assessment of wetlands and their functions. This could lead to the selection of particular wetlands, in the light of their functions and characteristics, for rehabilitation, sustainable use planning, conservation or complete transformation. This would of course have to be undertaken in consultation with the communities.

International Partnership

In Ethiopia, the initiatives described above are being undertaken by two NGOs acting in partnership with technical support from a university. These are:

- Wetland Action, a European grouping of technical and research specialists (www.wetlandaction.org),
- Ethio Wetlands and Natural Resources Association a local NGO, and
- Centre for Wetlands, Environment and Livelihoods at the Huddersfield University (http://wetlands.hud.ac.uk).

These organisations have worked together for five years and are beneficiaries of support from the Convention to Combat Desertification and a number of bilateral donors in Ethiopia.

This work in Ethiopia is generating lessons for use in several other wetland areas in the country and also in other parts of the Nile Basin and beyond. Specifically Wetland Action has provided technical inputs into community management of wetlands in the Nile Headwaters in Rwanda and in Zambezi headwater wetlands in Zambia.

References

Hailu, Ato Afework, Wood, A.P. and Dixon, A.B., 2003. Interest groups, local knowledge and community management of wetland agriculture in South-West Ethiopia. *International Journal of Ecology and Environmental Sciences,* **29:** 55-63.

Dixon, A.B., 2003. Indigenous Management of Wetlands: Experiences in Ethiopia. Ashgate, Aldershot.

Dixon, A.B. and Wood, A.P., 2003. Wetland cultivation and hydrological management in East Africa: Matching community and hydrological needs through sustainable wetland use. *Natural Resources Forum,* **27(2):** 117-129.

Dixon, A.B., 2002. The hydrological impacts and sustainability of wetland drainage cultivation in Illubabor, Ethiopia. *Land Degradation and Development,* **13(1):** 17-31.

Dixon, A.B., 2001. Indigenous hydrological knowledge in South-Western Ethiopia. *Indigenous Knowledge and Development Monitor,* **9(3):** 3-5.

Howard, G.W., Bakema, R. and Wood, A.P., 2005. The multiple use of wetlands in Africa. *In:* Maltby, E. (Ed.) The Wetlands Handbook. Macmillan, London (in press).

Wood, A.P. and Dixon, A.B. (Eds), 2001. Sustainable Wetland Management in Illubabor Zone: Research Report Summaries. The University of Huddersfield and Wetland Action, Huddersfield.

Wood, A.P., 2003. Sustainable wetland use, food security and river flow: Some experiences from the highlands of Ethiopia and Rwanda. *Revue de Geographie Alpine,* **91(2):** 57-68.

Wood, A.P., Afework, Hailu, Abbot, P.G. and Dixon, A.B., 2002. Sustainable management of wetlands in Ethiopia: Local knowledge versus government policy. *In:* Gawler, M. (Ed), Strategies for Wise Use of Wetlands: Best Practices in

Participatory Management: Proceedings of a Workshop held at the 2nd International Conference on Wetlands and Development, November 1998, Dakar, Senegal. IUCN, Gland, Switzerland.

Wood, A.P. and Dixon, A.B., 2001. Sustaining the Flow: Community Management of Highland Wetlands in South-West Ethiopia. *In:* Proceedings of the World Mountain Symposium, 30th September-4th October 2001, Interlaken, Switzerland (Online: http://www.wms2001.ch/cd/).

Wood, A.P., 2001. Gender and wetland development. *In:* Yilma Dellalegn (Ed.) *The Wetlands of Ethiopia. IUCN, Gland, Switzerland.*

9

Changing Flow in the Okavango Basin: Upstream Developments and Downstream Effects

J. Wilk, L. Andersson[1], P. Wolski[2], D. Kgathi[2], S. Ringrose[2] and C. Vanderpost[2]

Department of Water and Environmental Studies, Linköping University
S-581 83 Linköping, Sweden

[1]Swedish Meteorological and Hydrological Institute (SMHI)
S-601 76 Norrköping, Sweden

[2]Harry Oppenheimer Okavango Research Centre (HOORC)
Maun, Botswana

Introduction

The Okavango basin is shared between the countries of Angola, Namibia and Botswana and terminates in the Kalahari MOZ (Makgadigadi-Okavango-Zambezi) depression as an extensive alluvial fan often referred to as a "Delta" (Ringrose et al., 2005a). The upper catchment area receives 1200 mm/year rainfall and flows to the semi-arid Kalahari where the nominal average 460 mm/yr is considered a good rainfall year. Most of the streamflow in the basin is generated within the Angolan upper catchment. After 27 years of civil war, the cease-fire in 2002 may ultimately result in large number of refugees returning to the Angolan headstreams area with anticipated increased demands both for irrigation water and sites for dam construction for electricity generation. As the level of development is not high most of the returning people will be dependent on natural resources. Though the provision of needs to basin inhabitants is undisputed, there are concerns that the resettlement of displaced communities might have downstream environmental impacts (Green Cross International, 2000). Development will however be slow because of the large number of remaining landmines (Mendelsohn and El Obeid, 2004).

The downstream sections of the basin in Namibia and Botswana suffer water scarcity both in terms of human needs and in terms of the sustenance of valuable ecosystems. The demand for basin water may be further threatened by increasing demands from outside the basin, and supply could be much reduced as a result of climatic change (Andersson et al., manuscript, Wolski et al., manuscript). Stakeholders in Botswana for instance are concerned about how a planned potential pipeline from the Okavango river to central Namibia (mainly Windhoek) will impact ecosystems and hence tourism developments in the Okavango Delta.

Model-derived results can create a common view on the problems and serve as a basis for discussions about choices of strategies to avoid unwanted consequences of human use of natural resources. This holds true particularly for such climatically and culturally diverse basins as the Okavango. Models can also assist in dialogues of how to adapt to future problems that are beyond regional control, such as decreasing water availability due to changing climate (Schultze et al., 2004). During the European Union funded WERRD (Water for Ecosystem Resources and Rural Development) project, actual and possible future fluctuations of hydrological variables have been assessed (Andersson et al., manuscript, Wolski et al., manuscript) and their potential impacts downstream in terms of ecosystem responses (Murray-Hudson et al., manuscript). The scenarios include factors such as climate change, irrigation, dam construction, and water abstractions and their influence is weighed against the impact of climatic variability. Changes in hydrological inputs to the flood-pulsed Delta manifest themselves by changes in such flooding characteristics as water depth, inundation duration, and when considered on the long-term basis (decadal), in flooding frequency. In addition the human dimensions of downstream change have been articulated in terms of linkages between natural resource usage and socio-economic dynamics and policy (Kgathi et al., manuscript). The extent to which these changes will compare with past shocks to the Delta is not entirely known but it is possible that they will be much larger (especially the effects of climate change) than what the inhabitants have experienced up to now.

The aim of this work is to further improve the understanding of the dynamic linkages between human and riverine water resources in the Okavango River system by analysing more closely the impact of changes in the headwaters of the basin on water and ecosystem resources downstream and the effect on Delta livelihoods.

Methods

Finally, the fact that model-derived results only are useful if they are understood, accepted and perceived as legitimate by those concerned is discussed, and a way forward in this direction is suggested.

Landuse Cover

Sixteen late wet-season Landsat ETM+ images for 2003 were obtained in order to develop a landuse-cover baseline assessment. This was an advantage as it reflected the latest possible overview of the land-cover/land-use situation in the basin in view of ongoing post-civil-war resettlement of displaced persons along the Okavango tributaries in Angola. Image files were imported using Erdas-Image software routines and were re-projected to geographic coordinates (latitude/longitude) using the Erdas Reproject function to facilitate the bridging of several UTM map projection zones in the basin.

After experimenting with both unsupervised and supervised classification routines in Erdas Image, classification was performed on single images (as opposed to mosaics) to avoid inter-image reflectance contrasts and resulting misclassification (cf. Ringrose et al., 1988). Subsets of single images were re-created to cover the same 10 kilometres zone along the rivers and tributaries thereby focusing the classification on key elements in the immediate river zone. Preliminary statistical data in terms of prevalent cover percentages were developed for three landcover classes (riparian zones, cleared areas and irrigated farms) regarded as being of paramount importance in terms of their potential effects on river hydrology. Previous work has shown that the riparian trees are often deep-rooted thereby protecting the banks from erosion (cf. Ringrose et al., 2003). Procedures were thus carried out to determine the relative proportions of cleared land, irrigated land and riparian forest cover in the 10-km buffer zones in the respective image scenes.

Modelling of the Inflow to the Delta under Various Scenarios

A monthly rainfall-runoff model, divided into 24 sub-basins of which 18 had flow records of varying length at their outlets, was applied (Figure 1). It is a modified version of the Pitman model, including more explicit ground and

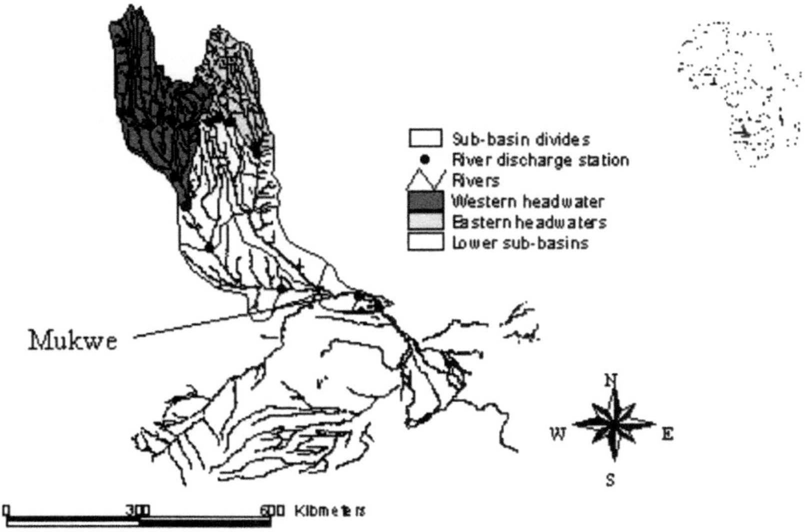

Figure 1: Sub-basins of the Okavango that contributes with inflow to the Delta. Locations of streamflow gauging stations are shown.

surface water interactions (Hughes et al., manuscript). A channel routing function was added, and riparian areas that are fed by seepage from the river, and from which water was assumed to evaporate, were modelled as open water surfaces.

In the Angolan part of the basin, where most of the streamflow is generated, gauged streamflow and precipitation data was only available between 1960 and 1972. To extend the rainfall database into more recent times, a diurnally corrected monthly dataset was created using passive and active microwave satellite data from the Special Sensor Microwave Imager (SSM/I) at monthly timescales and at 0.5° spatial resolution. Data from the Tropical Rainfall Monitoring Mission (TRMM) were used to overcome the diurnal bias in the SSM/I data (Wilk et al., manuscript). In order to extend the use of the model from the 1960-1972 period to the 1991-1997 period, it was necessary to ensure that the satellite rainfall data had the same basic characteristics as the gauged rainfall data. A correction equation was developed by comparing the gauged and original satellite rainfall frequency curves for three sub-basins and manually fitting a simple power function (Wilk et al., manuscript).

The complete period (1960-1972) for which historical rainfall data were available was used for calibration and is dominated by moderate to wet years. The period 1991-1997 for which satellite derived rainfall data were available, and for which streamflow records only were available for the downstream stations of Rundu and Mukwe in Namibia and Mohembo in Botswana, represents a drier period. This period was used to ensure that the calibrated parameter values were appropriate for use in future water resource estimations when the model is driven by satellite derived rainfall data, which will probably be the only available source of rainfall information in the foreseeable future.

Development scenarios were partly set up in co-operation with OKACOM representatives from Angola, Botswana and Namibia. Scenarios included one set of assumptions, based on "business as usual", and one based on "high development". Factors included were population and livestock increases, small-scale irrigation, larger irrigation schemes, deforestation, and construction of dams (Andersson et al., manuscript). Image processing of land-use/cover in the upper Okavango catchment resulted in nine classes (Wilk et al., manuscript). Estimates of precipitation (P) and potential evaporation (ET) from long-integration of the various global climatic change scenarios under two greenhouse gas emission scenarios were obtained from the IPCC data distribution centre. The data were averaged over all grid cells whose centres lie within the Okavango basin region. For each model run we defined a monthly 'delta' factor for the 30-year mean. The delta factor is the particular quantity during the middle (2020-50), late (2050-80) and end (2070-99) of the 21st century, relative to the 'present day' baseline period (1960-90) (Andersson et al., manuscript).

Modelling of the Hydrology of the Okavango Delta under Different Scenarios

In view of the spatial complexity of the system and paucity of data required for spatially distributed parameterisation of hydrological processes, the semi-distributed semi-conceptual approach, based on large units, i.e. a reservoir model, has been used. For the downscaling of the semi-lumped characteristic of flooding obtained from the reservoir model into the spatially-distributed characteristics of flooding necessary for ecological assessments, a GIS model has been developed. The details of the model are described by Wolski et al. (manuscript).

The reservoir model developed in this study simulates the Okavango Delta as a set of eight inter-linked reservoirs representing the major distributaries of the Delta. In the model, an upstream reservoir has separate outlets feeding downstream reservoirs. More than one outlet can provide a link between two reservoirs. In this way the non-linear character of the flow could be simulated. Infiltration and groundwater flow processes are explicitly represented by a set of reservoirs. Each of the surface water units has been divided into five sub-units represented by independent groundwater reservoir sets, comprising two reservoirs each: a floodplain groundwater reservoir and an island groundwater reservoir. Some model parameters (such as areas of reservoirs and topographic parameters) have been derived from the available satellite images, land cover maps (McCarthy and Gumbricht, 2004) and DEM (Gumbricht et al., 2004a). Flow parameters, due to their conceptual character, were calibrated in the model. Calibration of the model targeted the most reliable measurements of system response, namely discharges observed at the outlet of one of the distributaries and flood sizes observed for each of the units, as well as for the entire system. Calibration was done by trial-and-error adjustments of model parameters.

A hydrological database of the Okavango delta was consolidated whereby data on surface water levels and discharges were collected from DWA (Department of Water Affairs) and various published and unpublished quality controlled sources. The database was used to calibrate the model to represent most of the observed variability of system responses. Statistical and descriptive comparisons of hydrometric data and flood extent data for assessing the usefulness of hydrometric data in calibration of the regional reservoir model were done.

A method of translating (downscaling) the semi-lumped characteristics of flooding obtained from the reservoir model into the spatially-distributed characteristics of flooding necessary for ecological assessments was also developed.

The model system was used for assessments of the impact on hydrological characteristics of the Delta under various development and climatic change headwater scenarios. For each scenario, the extent of functional hydro-ecological units, and characteristics such as flood duration or flood frequency was determined and compared with that obtained under reference conditions.

Livelihood Study as Related to Risks and Shocks in the Downstream Delta

Primary data was collected from five study areas (Sehitwa, Seronga, Etsha, Gudigwa and Shorobe) within the Delta area through formal questionnaire interviews and focus group discussions. The formal interview questionnaire survey was undertaken in 2003, after doing a pre-test in the village of Tubu in the western part of Ngamiland. The sampling unit was a household, and those interviewed were heads of households. The number of households interviewed in each village was 36 in Etsha 6, 18 in Gudigwa, 28 in Sehitwa, 30 in Seronga, and 17 in Shorobe, giving a total of 129 in all the villages. The households were categorised into poor, middle, and rich peasants according to the number of assets they held (e.g. cattle, arable fields, gun, vehicle, type of house) and income earned. This information was cross-checked with information obtained from some key informants on the categorisation of households in Gudigwa and Shorobe, and there was a correlation in most cases. Data for the household questionnaire survey was analysed by calculating frequency distributions, measures of central tendency, and variation. Cross tabulations were used to determine the association between variables.

Focus group discussions were also held in three villages (Shorobe, Seronga and Gudigwa) where the groups were composed of six people each, from different parts of the village. The discussions included: livelihood activities in the village and their ranking, the drying of the river channels, the killing of cattle in order to eradicate the cattle lung disease, drought, damage of crops by wildlife, HIV/AIDS, and indicators of wealth.

Results

Landuse Cover

In general, a sharp contrast was noticeable between the northern (Angolan) and the southern (Namibian) river banks. All irrigated land and most cleared land is located along the southern bank in Namibia with very limited evidence of contemporary agricultural activity on the Angolan bank of the river, although old abandoned and overgrown (former) fields are in evidence. The extensive cleared lands on the Namibian side are a potential source for increased levels of sand deposition into the river channel, while the irrigation schemes have the potential to contribute chemicals deriving from fertiliser usage in addition to pesticides and herbicides in the future as the clearing and irrigated areas increase. This is particularly relevant to the area near Rundu, the largest town in the Kavango region.

The estimates of cleared, irrigated and riparian cover for the Angola portion of the river basin are based on comparative analysis with the Namibia section and known transitions to wetter, more vegetated conditions at lower latitudes in Angola. Large stretches of the Cuito river consist of extensive floodplains with the river meanders through these at a very slow pace. These floodplains are swampy and play an important role in regulating the flow of this major tributary of the Okavango. The Cuito is therefore a very balanced water contributor, delivering a more steady flow to the system than the Cubango itself, which is more prone to fluctuations. Along most of the Cuito, there is limited land clearing, except in the far north of the basin. Along the Cubango, land clearing for agricultural purposes is much more extensive especially around Menongue and Donde where the proportions reach 19.9% and 13.8% respectively. The upper headwaters especially occur in a densely populated region with large areas of fields cultivated mostly for subsistence farming, with some evidence of former large-scale plantations also visible on satellite imagery.

Model Results

With regard to the development scenarios in the upper catchment, the impact of most factors were rather limited, with a maximum impact on average annual flow due to population increase, deforestation irrigation, and other predicted possible future schemes of water abstractions being less than 5%. As the land-use study has shown, in the under-developed Angolan upper catchment there is relatively little land clearing at present and impacts on water use and quality are expected to be minimal. Much more extensive development, some of which involves irrigation and chemical usage, takes place in Namibia along the southern bank of the Kavango River. However, the impact was more pronounced for dry years, especially with regard to dams. For climate change scenarios, the impact varies considerably depending on the combination of global climate change model and greenhouse emission scenario used. The results indicate, however, that the possible impact of climatic change is larger than that of activities undertaken within the basin (Figure 2) (Andersson et al., manuscript).

The Delta climate is characterised by sequences of wetter and drier years, and has been subject to three general wetness regimes in the recent past: wet (1974-1985), transitional (1985-1990) and dry (1990-2000). These periods, particularly the wet and the dry, were of sufficient duration and consistency to result in the establishment of a particular distribution of wetland vegetation, and, to a lesser extent, woody vegetation. Given this situation, the assessment of future change was done separately for wet and for dry conditions. The effects of the change in hydrological inputs to the Delta as simulated by the scenarios, was therefore expressed by the change in size and distribution of hydro-ecological units that have varying flood duration and frequencies: permanent swamp (PF), regularly flooded seasonal floodplains (RFS),

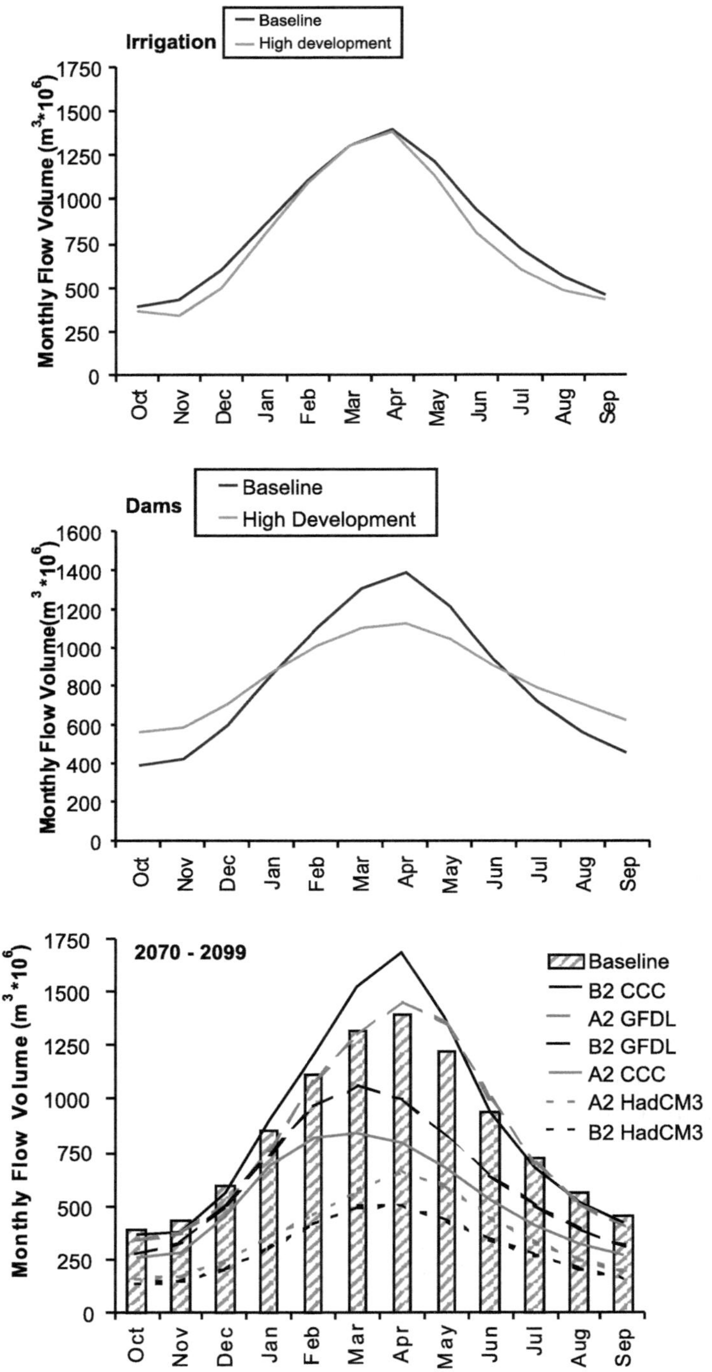

Figure 2: Changes of average monthly flow volume at Mukwe (Figure 1) from scenarios of irrigation and dam constructions under a high development scenario, and from various climate change scenarios.

occasionally flooded seasonal (OFS), high floods only (HFO) and dryland (D) (Table 1). Details of the procedure are described by Murray-Hudson et al. (manuscript).

Table 1: Floodplain classification with description of flood frequency and flood duration (months/year)

Floodplain class	*Sub-class*	*Class code*	*Flood frequency*	*Flood duration (months/year)*
Permanent floodplain	Proper	PFp	1	12
	Fringe	PFf	1	8-12
Regularly flooded seasonal floodplain		RFS1	1	4-8
		RFS2	0.5-1	
Occasionally flooded seasonal		OFS	0.1-0.5	1-4
High floods only		HFO	<0.1	<2
Dryland		DL	0	0

Simulations showed that upstream abstractions for human consumption are likely to have small short-term effects on the flooding pattern in the Delta, while other developments such as damming or deforestation have more pronounced effects. Figure 3 shows that the impact on water inflow to the Delta from the development scenarios in the upper basin (construction of dams and construction of all planned irrigation schemes) were shown to have consequences of some (usually small) magnitude to the hydrological characteristics of the entire Delta (as compared to the baseline). All of these effects are small, however, when compared to the changes resulting from existing climatic variability, and from the possible effects of climate change.

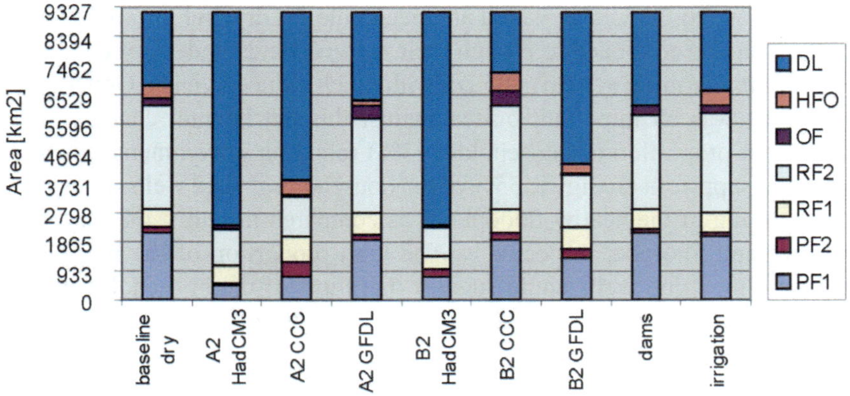

Figure 3: Simulated cumulative flood plains area (km^2) according to various development and climate change scenarios.

The combined effects of human abstraction and climate change as manifested by decreased rainfall and increased evaporation may result in significant Delta drying but some climate change scenarios also indicated increased flooding. The simulated changes resulting in changed patterns of Delta ecosystems will have varied effects on ecological processes in the Delta and relationships with the surrounding woodland and savannah. Additional ripple effects on trophic levels can be expected, for example as floodplain grasslands are succeeded by woodland in response to decreased flooding. These may also have downstream impacts on tourist facilities presently on the fringes of the seasonal swamps. Also, some effects are anticipated with respect to the aquifer recharge, which are both used for drinking water and to sustain riparian cover.

Livelihood Impacts and Responses to the Drying of River Channels

The main responses to past experiences of the drying of rivers (channels and floodplains) for the five villages were the drilling of boreholes, seeking government assistance and migration. In Sehitwa, the main strategies were: digging of wells (28%), switching from *molapo* (flood recession) farming to dryland farming (28%), and a growing dependence on the government for water supply (16%). A village group who caught, dried and sold fish to schools and health posts in the 1980s no longer exists after the river channels dried up. The household survey also revealed that crop yields were lower due to the switch from the more productive *molapo* farming to less productive dryland farming. In Shorobe focus group discussions revealed that the drying of the river channels in Shorobe was a disaster with even more dramatic impacts than the 1996 eradication of cattle in the area (to eliminate bovine pleuro-pneumonia). While only a few households owned cattle and those affected were partially compensated, desiccation of the river channels affected everyone, and no compensation was available. Because of the sandy and infertile soils in the village which are unsuitable for dryland farming most of the households continued to plough their *molapo* fields under dry conditions (dry *molapo* farming system), though with much reduced yields. Government assistance was an important buffer against this detrimental change, and a significant proportion of households (43%) relied on government assistance for water supply. In Gudigwa, 55% of the households drilled wells in response to the drying of the nearby floodplains as compared to only 14% and 5% in Sehitwa and Shorobe, respectively. The high proportion of the households who said they dug wells could suggest that most of the households did not have them in the village, as they had just been resettled in Gudigwa from different areas.

Another common response in the villages to the drying of river channels and floodplains was internal migration and local movements. In the study villages, 27% of the households reported that some people migrated or moved to other areas after the river dessication. The most striking movements, both in the form of internal migration and short-term mobility, occurred in Shorobe, Sehitwa, and Etsha 6. In Shorobe, 59% of the households interviewed revealed that the drying of the Gomoti river caused some of the households to migrate locally as their shallow wells dried. The drying of the Gomoti floodplains resulted in some people emigrating from a number of localities close to the village of Shorobe, which had water resources supplied by the government, and to other nearby villages. Our informal interviews revealed that those who remained in the above localities when others emigrated to Shorobe were mainly the poor households, as they could not afford the costs of relocation. On the other hand, those who emigrated to Maun from Shorobe were mainly the economically active. In Sehitwa, 54% of the households reported that some people emigrated to other areas where ground water or surface water could easily be obtained. It was also reported that others moved their livestock to other areas, including those near the buffalo fence where there was plenty of ground water. This pattern is consistent with a number of studies on rural settlement in Botswana which have revealed (Silitshena, 1980) that the availability of water is one of the reasons why people do not permanently settle. Lack of water is a factor that induces people to leave rural areas while the villages attract them with employment, schools and clinics.

Benefits Arising from the Project and Recommendations for Future Work

Landuse cover data has been established over the Okavango basin but more work is needed showing the impact of change both in terms of out migration during the civil war and the impacts of the more recent in migrational trends. Hydroclimatological databases, and other databases relevant for hydrological simulations have been constructed for the Okavango basin, and a hydrological model has been set up. The results indicate that the model more than adequately represents the hydrological response of the basin. Consequently, it can be used to assess the impact of future climatic and development scenarios. The fact that the model has performed satisfactorily across a range of historical climatic conditions (wet and dry periods) suggests, that the calibrated model can also be used to assess flow regimes under various climate change scenarios. For future work, confirmation against more detailed hydroclimatological and physiographic data from the basin is recommended. Also for better validation of the satellite dataset in the Okavango region, more rain gauge measurements are needed. Consequently, without more effort being spent on monitoring of the basin, it cannot be expected that model performance or validity can be significantly improved. The performed remote sensing classifications of land use in the basin revealed that much more extensive development presently takes place in Namibia, compared to in Angola, but for some parts of the basin a similar development might be seen in the future also in Angola. However, the most immediate need is towards monitoring of the impact along the southern bank of the Okavango river in Namibia, which are likely to be the most detrimental mainly from the perspective of water quality declines, in terms of immediate potential impacts on the Okavango Delta.

For further work on the quantification of channel losses, more work is recommended on combining the modelling approach and the remote sensing derived landscape information. Use of remote sensing derived data for better spatial and temporal estimates of evaporation could also be beneficial. Scenarios have been evaluated with regard to the impact of developments within the basin and climatic change. Evidently, stakeholders and decision-makers in the basin might see other scenarios than those presented as more relevant. A clearer connection between socio-economic driving factors in the region under development should be incorporated in any future model-based applications. It is therefore necessary to build up capacity in the region in the use of models to assess hydrological consequences of various scenarios. This would facilitate OKACOM (Okavango Basin Commission) members in stakeholder dialogues and in their work on the Integrated Basin Development Plan.

The developed reservoir model of Delta flooding with its comprehensive nature, i.e. explicit simulation of floodplain and island groundwater, as opposed to the previous models (Wolski and Gumbricht, manuscript), made it possible to simulate the effect of the 1974-1981 "high flow regime" of Boro. Additionally, the model confirms and provides more insight into the presence of regional non-stationarities in flood distribution. The model also improves understanding of behaviour of the system at sub-regional scale, shedding some light on the relationship between channel and floodplain flows. In general, the prepared model seems to be the most reliable and realistic model of the system created so far.

The livelihood study has shown how people have reacted to the drying of channels in the past which gives indication of what options are available to people if and when this occurs again. Whether or not new conditions will affect their opportunities to react in the same way (e.g. will the existence of zoning and fencing limit people from moving as readily as they have done in the past) will remain to be seen. The effects of some of the simulated upstream changes if they occur are also of another magnitude and duration than past changes that will also influence how many options people will have. Some climate change scenarios even indicate increased flooding and new strategies will have to develop in order to adapt. One definite trend in times of hardship is the increasing dependence on government subsidies and other income-generating programmes. More work is needed to assess what options are available if even larger changes in flooding negatively affect ecosystems and livelihood bases and how mitigation strategies can be best implemented.

Acknowledgements

This study formed part of the EU funded project WERRD (Water and Ecosystem Resources in Regional Development – Balancing Societal Needs and Wants and Natural Systems Sustainability in International River Basin Systems). Contributions, in terms of hydroclimatological data and information on the basin characteristics, are gratefully acknowledged from the Botswana and Namibian Departments of Water Affairs and Departments of Meteorological Services, Steve Crerar (Okavango Basin Preparatory Assessment Study), John Mendelsohn, Selma el Obeid, FAO, USGS, the Land Use and Land Cover Change International Project Office, Miombo Network, the Global Vegetation Modelling Unit, the Tyndal Centre, as well as from other team members of WERRD. We also thank Carlos Eduardo Cabrel Carvalho for compiling the databases and quality checking the data and Mr A.C. Diniz and Mr J. Rochas from LNEC in Lisbon for discussions about the feasilibity of irrigation schemes in Angola.

References

Andersson, L., Wilk, J., Hughes, D., Kniveton, D., Layberry, R., Todd, M. and Savenije, H.H.G. Scenarios of the impact of changes of climate and water use on water flow in the Okavango River. *Manuscript.*

Green Cross International, 2000. The Okavango River Basin. Transboundary Basin Sub Projects: The Okavango. From: www.gci.ch/GreenCrossPrograms/waterres/pdf/WFP_Okavango.pdf.

Hughes, D., Andersson, L., Wilk, J. and Savenije H. Regional Regional calibration of the Pitman model for the Okavango River. *Manuscript.*

Mendelsohn, J. and el Obeid, S., 2003. Sand and Water: A Profile of the Kavango Region. Struik Publishers, Cape Town, South Africa.

Kgathi, D.L., Ngwenya, B. and Wilk, J., 2005. Impact of shocks on rural livelihoods and household response: The case of Ngamiland in Botswana. Submitted to *Special Edition of Ecology & Society.*

Murray-Hudson, M., Wolski, P. and Ringrose, S. Scenarios of the impact of local and upstream changes of climate and water use on water flow in the Okavango Delta. *Manuscript.*

Ringrose S., Huntsman-Mapila P., Kampunzu, A.B., Downey, W.D., Coetzee, S., Vink, B., Matheson, W. and Vanderpost, C., 2005a. Geomorphological and geochemical evidence for palaeo feature formation in the northern Makgadikgadi sub-basin, Botswana. *Palaeogeography, Palaeoclimatology & Palaeoecology,* **217(3-4):** 265-287.

Ringrose, S., Jellema, A., Huntsman-Mapila, P., Baker, L., Brubaker, K., 2005. Use of remotely sensed data to assess the significance of drying gradients in the Okavango Delta, Botswana. *International Journal of Remote Sensing* (Accepted).

Ringrose, S., 2003. Characterisation of riparian woodlands and their potential water loss in the distal Okavango Delta, Botswana. *Applied Geography,* **32(2):** 281-302.

Ringrose, S., Matheson, W. and Boyle T., 1988. Differentiation of Ecological Zones in the Okavango delta, Botswana by classification and contextual analysis of Landsat MSS data. *Photogrammetric Engineering & Remote Sensing,* **54:** 325-332.

Schulze, R., Horan, M., Seetal, A. and Schmidt, E., 2004. Roles and Perspectives of the Policy maker, Affected Water Sector and Scientists in Integrated Water Resource Management: A Case Study from Africa. *Int. J. Water Resour. Dev.,* **20:** 325-344.

Silitshena, R.M.K., 1980. Population movements and settlement patterns in contemporary Botswana. *In:* Proceedings of the Symposium on Settlement in Botswana: the Historical Development of a Human Landscape. Heinmann Educational Books, Gaborone, Botswana.

Wilk, J., Kniveton, D., Andersson, L., Ringrose, S., Layberry, R., Todd, M, Hughes, D. and Vanderpost, C. Rainfall, water balance and land use of the Okavango basin upstream the delta. *Manuscript.*

Wolski, P. and Gumbricht, T. Modelling of the hydrology of the Okavango delta. *Manuscript.*

Wolski, P, Ringrose, S. and Vanderpost, C. Scenarios of the impact of local and upstream changes of climate and water use on water flow in the Okavango Delta. *Manuscript.*

10

Bedrock Groundwater as a Major Control on Streamflow Generation in Upland Wales, UK

Atul H. Haria and Paul Shand[1]
Centre for Ecology and Hydrology (formerly Institute of Hydrology)
Maclean Building, Crowmarsh Gifford
Wallingford, Oxfordshire OX10 8BB, UK
[1]British Geological Survey, Maclean Building, Crowmarsh Gifford
Wallingford, Oxfordshire OX10 8BB, UK

Introduction

Hard rock upland regions source some of the largest rivers in the UK and therefore constitute an important water resource. The water quality of these headwater streams has been the subject of concern for a number of decades; the predominant focus in NW Europe has been in respect to catchment acidification and the resulting impact on stream ecology (e.g. Harriman and Morrison, 1982). The stream water quality response of these catchments to anthropogenic acid deposition is a function of the range of travel time distributions, resulting from differences in flow pathways, from when rainfall enters the catchment to its appearance in the stream channel (Kirchner et al., 2000, 2001). Understanding these flow pathways and the mechanisms of streamflow generation is fundamental for understanding the catchment response to pollution.

A number of studies of upland catchments have highlighted the dominance of pre-event water in the rapid stream response to rainfall (Sklash and Farvolden, 1979; Soulsby, 1995; Shand et al., 2004). To date, the source of this pre-event water and the mechanisms by which it enters the stream channel have not been explained. Traditional concepts have always maintained these upland catchments as being sealed below the soils where the bedrock geology was considered impermeable. However, hydrochemical studies from the Plynlimon catchments (Neal et al., 1988; Neal and Rosier, 1990; Shand et al., 1997) have produced data that do not fit the hypothesis of rapid stream response to rainfall generated solely by runoff over and through the soils. These hydrochemical data showed a damping of the rainfall chloride signal (Neal et al., 1988) indicating the presence of a reservoir of water or "mixing-pot" within the catchment where the signal is buffered prior to reaching the stream channel. Subsequent work in the catchment identified a significant groundwater reservoir in the bedrock geology (Shand et al., 1997; Neal et al., 1997) that has previously been ignored. However, the role of this reservoir and its contribution to streamflow generation processes remained unclear.

This paper presents some of the results from a comprehensive physico-chemical study in the headwaters of the Afon Hafren (River Severn) in mid-Wales, UK, designed to identify the role of bedrock groundwaters in streamflow generation. This work is driven by the need to understand the processes that determine catchment water resources and control the water quality of both surface and groundwaters. An understanding of these upland processes is, therefore, crucial for developing a sustainable aquatic ecosystem, fisheries and potable water resource.

Methods

The study site was located on the south-west facing slope of the Afon Hafren (Figure 1), in the forested Plynlimon catchment in mid-Wales, UK. The soil profile at the site is a stagnopodzol overlying a bedrock geology of Ordovician and Silurian shales and mudstones. A transect of four nests of instruments (the Hafren Transect) running perpendicular to the stream channel from the river bank to the lower hillslopes (nests A-D in a plot ca. 50 m × 10 m) was installed to observe soil, stream and groundwater processes (Figure 2). The aim was to monitor the physico-chemical changes in each of these hydrological

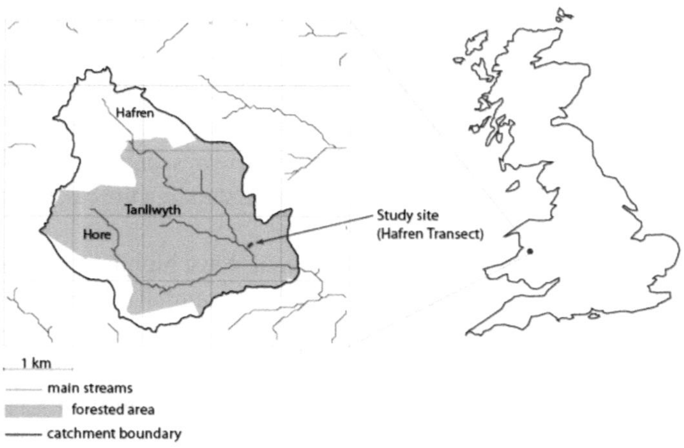

Figure 1: Location of the Hafren Transect study site in the forested Hafren catchment (Plynlimon, mid-Wales, UK).

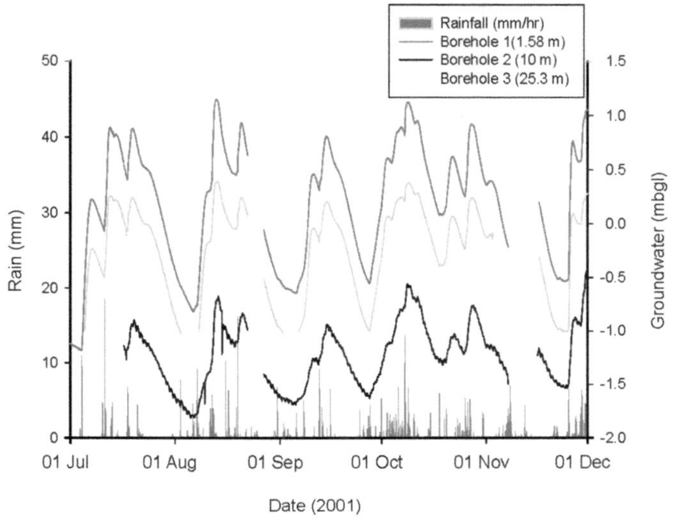

Figure 2: Schematic cross-section of the Hafren Transect study site.

compartments in response to rainfall events in order to understand flow pathways and mixing processes. Soil water processes were monitored using logging Pressure Transducer Tensiometers (PTTs) installed at approximately 0.1 m, 0.25 m and 0.4 m depths at each of the four nests. Aquifer characterisation was assessed using boreholes (along the same transect) that were sealed to specific depths using novel techniques in order to isolate and sample specific horizons and therefore any groundwater stratification.

Groundwater processes were monitored at approximately 1.5 m, 10 m and 25 m depths (representing the weathered, less weathered and consolidated bedrock respectively) in each of the four nests (Figure 2) by pressure transducers. All data were logged hourly. Groundwater samples were routinely taken from all boreholes (using submersible pumps), soils and the stream for chemical analysis. Selected results from the chemical monitoring are presented here; more details can be found in Shand et al. (2004). Further details on the site and instrumentation are given by Haria and Shand (2004).

Results and Discussion

Typical time-series groundwater level responses measured in the boreholes are shown for Nest D in Figure 3. Two things are of interest. First, the pressure heads do not increase or decrease in order with depth; the greatest head is in the intermediate horizon (BH2; ca. 10 m depth) indicating confined or semi-confined groundwater bodies in depth specific fractures at this site close to the stream channel. Pressure heads measured in some of the boreholes were artesian (e.g. BH2, Figure 3); however as this site close to the stream represents a groundwater discharge zone, this in itself does not mean confining conditions.

Secondly, despite the deeper groundwater bodies (ca. 10 m and 25 m depths) being confined, the pressure responses observed at all depths were very similar (Figure 3). This indicates that all three groundwater horizons were being recharged simultaneously somewhere upslope; the mechanism of recharge, however, still remains unclear. These data highlight an active hydrogeological system to at least a 30 m depth.

The groundwater stratification in physical parameters observed was further defined in the groundwater chemistry (Figure 4). Groundwater in the fractures in the shallower bedrock horizons (< 10 m) were acidic and oxidising, whilst the deeper fractures yielded groundwaters that were alkaline and reducing. CFC ages of the discrete groundwater compartments, combined with seasonal

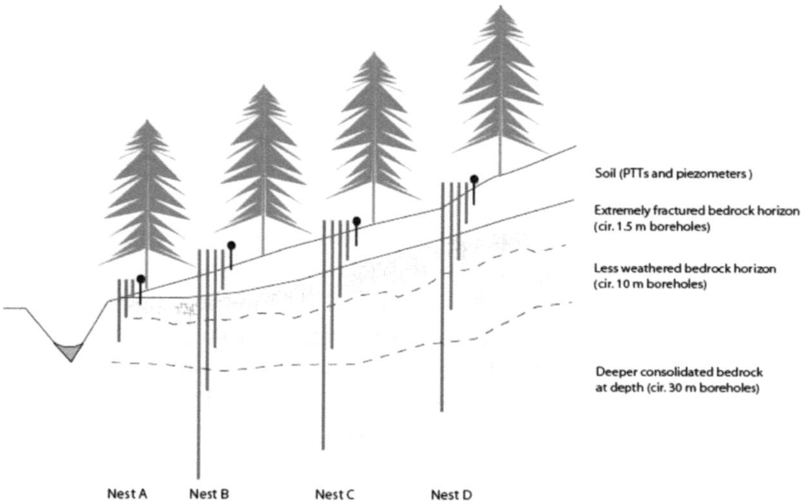

Figure 3: Groundwater heads measured in the boreholes at Nest D.

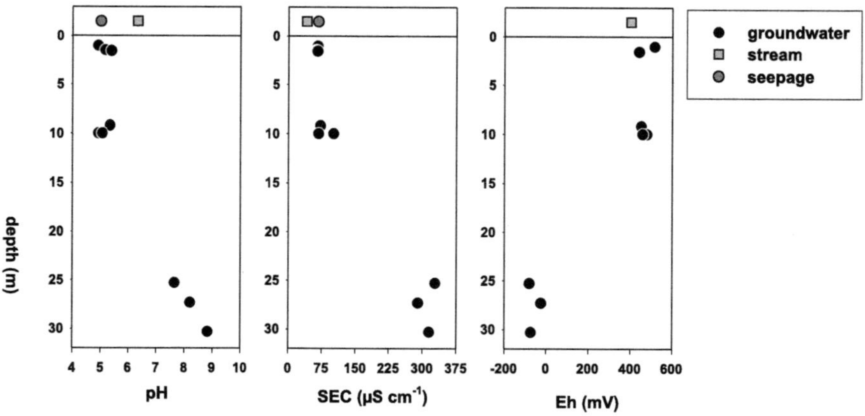

Figure 4: Variations in some groundwater parameters with depth.

stable isotope data, indicated a degree of mixing between shallower (ca. 1-10 m depth) younger groundwaters and deeper (ca. 30 m depth) older groundwaters.

The identification of an active but compartmentalised groundwater system in the bedrock geology leads to the question: *Is this groundwater contributing to streamflow? And if so, then when and by how much?* These questions are of further significance with respect to water quality when considering the extremes in water chemistry variation with depth.

The contribution of groundwater to streamflow is likely to vary from the different groundwater compartments. Groundwater head changes measured in the shallow 1.5 m boreholes over a storm event (10-12 July 2001) highlighted a negligible groundwater rise close to the stream whilst the degree of groundwater response increased with distance away from the stream (Figure 5). The "fixed" head close to the stream indicates a point of groundwater discharge as bank seepage. This bank seepage, previously mistaken to be new event water flowing through the soils, was shown to have similar hydrochemical characteristics to the shallow 1.5 m groundwater (Figure 4). Evidence of groundwater discharge to the stream channel from the intermediate 10 m system is presented by Haria and Shand (2004); CFC data showed a contribution of deeper (30 m) groundwater through various mixing processes (Shand et al., 2004).

The exact mechanisms and flow pathways involved have been described in detail by Haria and Shand (2004); for example, shallow 1.5 m groundwater storm event discharge was enhanced at the interface between the soils and the geology (ca. 40 cm depth) along a lateral fast flow pathway. This 40 cm depth also coincided with the horizon from where the greatest concentrations of Al were mobilised during storm response (data not shown). The postulated hypothesis for toxic soil contaminants reaching the stream channel quickly

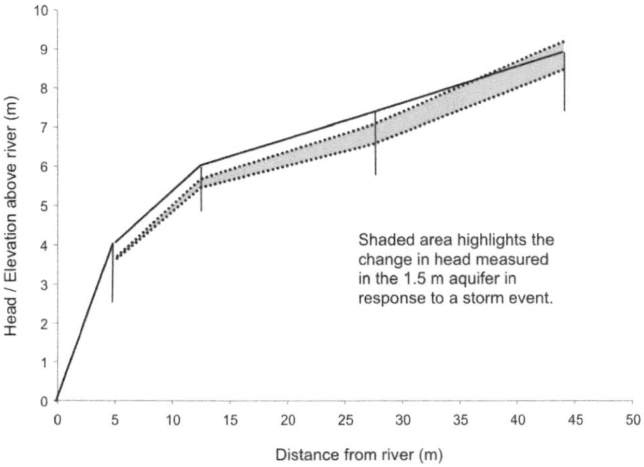

Figure 5: Cross-section showing groundwater head changes in the shallow 1.5 m aquifer over a storm event (10-12 July 2001).

during storm response presented in this paper is, therefore, not a fast runoff response from above, but a rapid piston displacement of pre-event groundwater from below up into the lower soils and then laterally in to the stream.

How much groundwater is reaching the stream channel varies from each sub-surface compartment. However work by Shand et al. (2004) at this site using δ^2H showed that the peak stream response to rainfall was made up predominantly of pre-event water (> 90%). These results agree with findings for other upland catchments (e.g. Sklash and Farvolden, 1979; Pearce et al., 1986; Soulsby, 1995).

Conclusions

This paper presents the findings from a detailed physico-chemical process-based field study looking at the storm event responses in soil waters, groundwaters and stream. The integrated nature of the study has highlighted interactions between different components of the water cycle taking into account the deep bedrock geology that has traditionally been (and to a large degree still is) considered impermeable.

Despite the evidence of deep bedrock groundwater (Neal et al., 1997; Shand et al., 2001; Haria and Shand, 2004), many hillslope hydrologists still use the term groundwater to mean rainfall derived saturation in the soil horizons; this often implies a perched and/or ephemeral system. The concept of a deeper bedrock groundwater resource (extending down to many 10s of metres depth) has been poorly understood and is consequently often ignored. Whilst saturation in soils can technically be termed groundwater, in this paper the term groundwater refers specifically to the permanent saturated horizons in the bedrock geology. Because bedrock groundwaters are older and chemically different to new event waters, bedrock groundwater rise into the soils causing saturation is wholly different to rainfall-derived perched saturated horizons in the soils.

The main results from this study are outlined below:

- Groundwater is present in the bedrock geology in distinct confined or semi-confined compartments.
- The groundwaters show a strong chemical stratification reflecting the compartmentalised system.
- Mixing occurs between the different groundwater compartments including the deeper waters from at least ca. 30 m depth.
- Groundwaters respond rapidly to rainfall and discharge, in different amounts from the different compartments, into the stream channel.
- The mechanism of groundwater discharge is by piston displacement in response to recharge further upslope.
- By this mechanism, upwelling shallow groundwater can rise into the lower soils and quickly transport toxic soil components (such as aluminium) along fast lateral flow paths to the stream channel.

- Groundwater is a major component of streamflow during peak storm response.
- Groundwater is a major control on surface water quality during storm response.

These and other process studies are beginning to delineate the importance of bedrock groundwater in stream flow generation in headwater catchments around the world. This work highlights the complex interactions between different components of the water cycle emphasising the fact that these processes cannot be considered in isolation. It is essential that these developments in process understanding are used to constrain existing models, especially with respect to water quality. Although scaling up from site to the catchment scale poses significant problems for existing models, upland catchment water quality models that do not account for these bedrock hydrogeochemical processes are likely to be severely flawed.

References

Haria, A.H. and Shand, P., 2004. Evidence for deep sub-surface flow routing in forested upland Wales: Implications for contaminant transport and stream flow generation. *Hydrology and Earth System Sciences,* **8(3):** 334-344.

Harriman, R. and Morrison, B.R.S., 1982. Ecology of streams draining forested and non-forested catchments in an area of central Scotland subject to acid precipitation. *Hydrobiologia,* **88:** 251-263.

Kirchner, J.W., Feng, X. and Neal, C., 2000. Fractal stream chemistry and its implications for contaminant transport in catchments. *Nature,* **403:** 524-527.

Kirchner, J.W., Feng, X.H. and Neal, C., 2001. Catchment-scale advection and dispersion as a mechanism for fractal scaling in stream tracer concentrations. *Journal of Hydrology,* **254:** 82-101.

Neal, C., Christophersen, N., Neale, R., Smith, C.J., Whitehead, P.J. and Reynolds, B., 1988. Chloride in precipitation and streamwater for the upland catchment of river Severn, mid-Wales; some consequences for hydrochemical models. *Hydrological Processes,* **2:** 155-165.

Neal, C. and Rosier, P.T.W., 1990. Chemical studies of chloride and stable oxygen isotopes in two conifer afforested and moorland sites in the British uplands. *Journal of Hydrology,* **115:** 269-283.

Neal, C., Robson, A.J., Shand, P., Edmunds, W.M., Dixon, A.J., Buckley, D.K., Hill, S., Harrow, M., Neal, M. and Reynolds, B., 1997. The occurrence of groundwater in the Lower Palaeozoic rocks of upland Central Wales. *Hydrology Earth System Sciences,* **1:** 3-18.

Pearce, A.J., Stewart, M.K. and Sklash, M.G., 1986. Storm runoff generation in humid headwater catchments 1. Where does the water come from? *Water Resources Research,* **22(8):** 1263-1272.

Shand, P., Darbyshire, D.P.F., Gooddy, D.C., Edmunds, W.M. and Neal, C., 1997. $^{87}Sr/^{86}Sr$ as an indicator of flowpaths and weathering rates in the Plynlimon catchments, Wales, UK. BIOGEOMON: The 3rd International Symposium on

Ecosystem Behaviour, *Journal of Conference Abstracts*, **2(2):** 296. Cambridge Publications.

Shand, P., Darbyshire, D.P.F., Haria, A.H., Davies, J. and Griffiths, K., 2004. Streamflow generation in upland impermeable catchments – is there a need to think deeper? *In:* Wanty and Seal II (Eds.) Proceedings of the Eleventh International Symposium on Water-Rock Interaction, Vol. 2, Saratoga Springs, New York, USA/A. A. Balkema Publishers, Netherlands.

Soulsby, C., 1995. Contrasts in storm event hydrochemistry in an acidic afforested catchment in upland Wales. *Journal of Hydrology,* **170:** 159-179.

Sklash, M.G. and Farvolden, R.N., 1979. The role of groundwater in storm runoff. *Journal of Hydrology,* **43:** 45-65.

11

CRENODAT (Biodiversity Assessment and Integrity Evaluation of Springs of Trentino (Italian Alps) and Long-term Ecological Research): Project Design and Preliminary Results

Marco Cantonati, Ermanno Bertuzzi and Alessia Scalfi
Museo Tridentino di Scienze Naturali
Limnology and Phycology Section, Via Calepina 14, 1-38122, Trento, Italy

Introduction

It is well known that springs are a fundamental source of good quality water and they frequently occupy a place of distinction in cultures and mythologies. On the contrary there is still very limited public awareness on the fact that they are also special habitats of great importance for nature conservation for a number of reasons recently reviewed in Cantonati et al. (2006). In limnology, together with streams and large lowland rivers, they are one of the main categories of running waters. Nevertheless, springs have been less studied than most other aquatic habitats.

In all the models of longitudinal zonation of running waters, from the "classical" (e.g. Illies and Botosaneanu, 1963) to the recent ones (Ward, 1994; Füreder et al., 2002), a specific zone is reserved for springs in recognition of the environmental peculiarities that characterise them. As pointed out in the "sandglass" model by Cantonati and Ortler (1998), the spring can be considered the lowest point of a subterranean hydrographic system and it is, at the same time, the origin of a surface hydrographic system. From an ecological point of view the most correct approach is probably to recognise this dual nature defining the spring as the ecotone linking surface with underground waters.

The most well known characteristic of springs is their environmental stability (Odum, 1971), that is the reduced amplitude of the seasonal fluctuations of the main physical and chemical factors. Springs have thus soon been proposed as a sort of "natural laboratories" (e.g. Round, 1981) where temperature is almost constant. Another characteristic of the spring habitat is the marked mosaic structure (e.g. Illies and Botosaneanu, 1963): in a few square metres a relevant number of microhabitats or choriotopes (e.g. Weigand, 1998) can be found. If one considers the number of taxa hosted by springs, these habitats turn out to be really worth to be protected (e.g. Cantonati and Ortler, 1998; Weigand, 1998). The frequently marked heterogeneity of environmental conditions and of communities and the presence of highly specialised taxa are important arguments supporting the conservation even of single environments.

Because of the small dimensions and of the importance of the fringing semi-aquatic habitats, springs are extremely sensitive to disturbance factors, such as trampling damage by cattle, coverage by sediment, water abstractions, stripping off of the surrounding vegetation, nutrient inputs etc. (Weigand, 1998). Springs are exposed to a series of direct (water abstractions and forest management activities) and indirect (by means of the deposition of nutrients or contaminants in the drainage basin) impacts. Among indirect impacts the following should further be mentioned: the deposition of different types of airborne contaminants in the drainage basin (acidification, increased N deposition, heavy metals), global warming, and increased UV radiation. The main threat for spring-habitats remains, however, their systematic use for drinking water supply, which implies the destruction of the original morphology and a severe impoverishment of the biota. In most intensely managed European areas springs habitats in near-natural conditions have virtually disappeared (e.g. Zollhöfer, 1997).

Researches specifically devoted to spring habitats are still rare and were carried out mainly since the beginning of the 1990s: Roca (1990), Sabater and Roca (1990, 1992), Cantonati (1998a), Crema et al. (1996), Haseke (1998), Zollhöfer (2000) and Werum (2001). Springs are habitats which are quite suitable for long-term monitoring since they integrate in space and time the effects of eventual direct and indirect impacts on the drainage basin (Cantonati et al., 2006). However, only in very rare cases, long-term research projects were started based on regional investigations on the springs of determined areas. In particular, two initiatives led to long-term ecological research projects in the Southern (Adamello-Brenta Natural Park; Cantonati and Ortler, 1998; Cantonati and Maiorana, 2001) and Northern (Berchtesgaden Natural Park, Bavaria, Germany; Gerecke et al., 2002) Alps respectively.

All groups of organisms can potentially provide useful information on the spring environment (cf. Cantonati et al., 2006). In the phytobenthos of mountain running waters – including springs—the dominant groups are generally diatoms and cyanobacteria (e.g. Cantonati et al., 2001). Diatom assemblages can provide precise information on the most relevant environmental factors to be considered in springs, such trophic status and geochemical preference (e.g. Rott et al., 1999), flow permanence (e.g. Werum, 2001) and lithology of the aquifers (Werum and Lange-Bertalot, 2004). Moreover, several authors (Kadlubowska, 1985; Cox, 1990; Cantonati, 1998b; Werum, 2001) report mostly reduced seasonal changes in the proportions of benthic diatoms in springs.

The goal of the present paper is to offer an outline of the CRENODAT Project and to discuss first results obtained during exploratory surveys on a group of large karstic springs.

Methods

The CRENODAT Project Proposal was submitted to the University and Scientific Research Service of the Autonomous Province of Trento for review and evaluation at the end of September 2003. The Project was officially approved by the local government on the 9th of July 2004 and activities were planned to last three years. In this first phase of the Project, special emphasis was laid in the selection of the study sites. In recognition of the importance of springs, in particular as potential sources of drinking water, several regional Administrations started inventory projects. The Autonomous Province of Trento began this activity in the 1970s. The database has been initiated with data collected in the frame of a project carried out in 1974-75. In 1993 a working group has begun the work of critical examination of the data to reach the objective of a single definition and labelling system of springs and to link to it the existing data and information. This database and field data files and photographs made available by the Geological Service of the Autonomous Province of Trento could be used for a first selection of 600 springs meeting the criteria for being considered within the Project CRENODAT. The most important ones were to include: mainly permanent springs; habitats not affected by heavy direct impacts; all lithologies of the study area (ranging from limestones and dolomites to schists and gneisses and to different types of granites); all altitude intervals (from less than 100 to more than 2500 m a.s.l.); and all districts of the Province with special attention to Biotopes and Natural Parks and to territories planned to be included in protected areas. This first list of possible sites will be integrated by means of expert interviews and considering also springs well known by the local population because of use for water supply since historical times, supposed curative effects of the water, or prominent impact on the landscape.

The suitability as project sites of as many springs as possible should be tested also by exploratory surveys. The first one was carried out already on the 20th of July 2004 and was devoted to a group of mostly large karstic springs emerging at the southern limit of the Brenta (Western Dolomites) mountain range, in the Municipality of Stenico. Four springs were sampled. The main physical and chemical parameters (morphology, substratum particle size, shading, temperature, current velocity, discharge) were measured or estimated in the field. Notes were taken on overall bryophyte abundance and on the occurrence of cyanobacteria and algae (other than diatoms) forming colourings or structures recognizable by the naked eye. More detailed sampling was devoted to diatoms: Epilithic material was obtained brushing about ten stones and epibryon was gathered squeezing bryophytes and adding also some bryophyte plants to the sample. These materials were prepared with hydrogen peroxide and hydrochloric acid to produce suspensions of cleaned valves and then permanent mounts using Naphrax®. At least 400 diatom valves were counted to obtain the proportions of the individual taxa. Shannon-Wiener Diversity Indices were calculated from these relative abundances.

To adopt the methods most suitable for the project goals a first workshop ("The spring habitat: biota and sampling methods") was organized in Trento at the beginning of March (28/02/05-03/03/05) with the involvement of the project specialists and consultants (Figure 1). Their contribution to the workshop are being collected to form a volume with the same title (Cantonati et al., 2007).

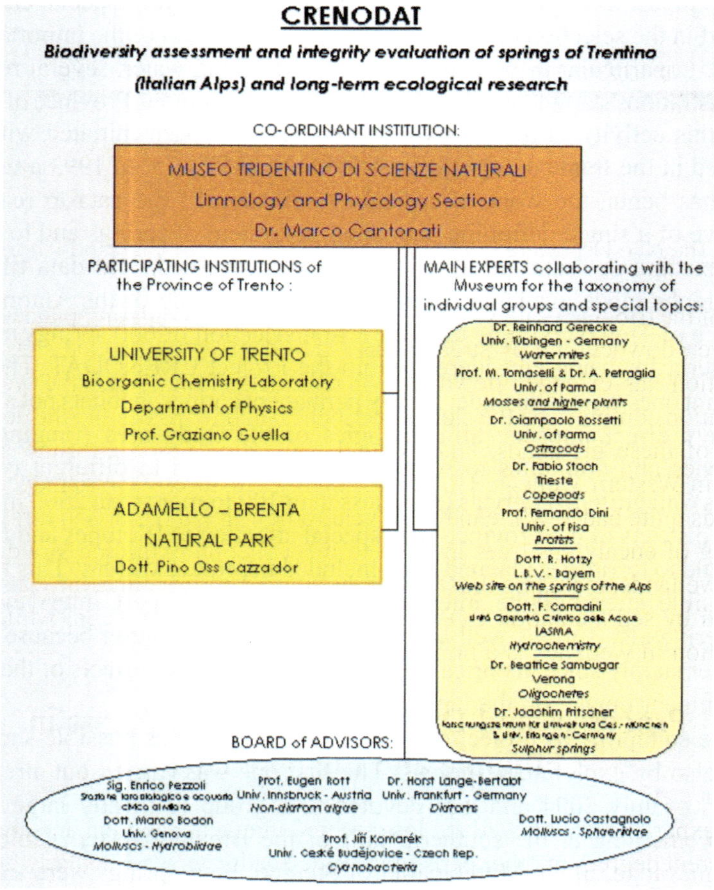

Figure 1: The organisation of first workshop in Trento.

Results

Background Information on the CRENODAT Project

CRENODAT aims at contributing to a thorough understanding of the structure of the biota; at obtaining a complete set of hydrochemical data covering the whole territory of the Province of Trento; at starting the investigation of the bioorganic metabolism of organisms fairly common and/or abundant in springs; at using springs for ecophysiological investigations and as ideal sites for long-term research; at selecting indicators of integrity/naturalness (community structure and typical taxa) and of environmental quality (organisms and bioorganic compounds) in relation to specific problems/parameters; and at promoting habitat-protection by means of pilot popularisation initiatives. The Project is coordinated by the Museo Tridentino di Scienze Naturali of Trento and carried out with two other local partners and thanks to the contribution of several international specialists and of a board of consultants (Figure 1). The Project acronym honours Gernot Bretschko, renowned limnologist who recently (2002) passed away. RITRODAT – a project devoted to the study of the functional ecology of a stream located near the Biological Station in Lunz (Austria) – was the long-term research dearest to this scientist.

For a more effective achievement of the project's general goal, activities were structured in the following six main tasks.

I: *Geomorphological and Hydrochemical Characterisation and Biodiversity Assessment*

A comprehensive study will be carried out on a representative sample of the springs of Trentino in near-natural conditions. Morphology, hydrogeology, chemical characteristics and the biota (algae, with special attention to diatoms, cyanobacteria, protozoa, water mites, microcrustaceans, crenobiontic and crenophilous snails, and other zoobenthos groups, bryophytes and vascular plants) will be investigated for about eighty springs representative of all main lithological substrata and in areas with relevant naturalistic quality. The analysis of several groups of organisms should allow to produce a complete picture of the biodiversity of these environments and to detect integrated environmental signals with the possibility of validation internal to the data-set. In collaboration with the Adamello-Brenta Natural Park, a pilot study for the detailed surveying of all springs of two particularly relevant areas of the Park, one on crystalline and one on carbonate substratum, will be organised. The exact knowledge of all the sources (location etc.) will further allow interesting future comparisons which should show the development of this part of the hydrographic system also in relation to climate changes.

II: *Springs with Special Physical and Chemical Characteristics*

The second task of the project focuses on about ten springs with special physical and chemical characteristics, such as thermal springs and/or sources with specific mineralization characteristics. Some sulphur springs will be included as well to further investigate potential detoxification capacities of the biofilms which thrive in these environments (Fritscher, 2004). Also the pointing out of biological indicators of specific types of waters (e.g. rich in heavy metals) will be a goal of this part of the Project.

III: *Springs as "Natural Laboratories" (Ecophysiology and Bioorganic Metabolism)*

The typical physico-chemical stability of springs will be exploited to use these environments as a sort of natural laboratory. Ecophysiological characteristics and the bioorganic metabolites of organisms thriving in great abundance in some springs (e.g. the chrysophyte *Hydrurus foetidus* Vill. (Trev.) and the diatom *Diatoma mesodon* (Ehrenberg) Kützing) will be studied. These investigations are hoped to be able to contribute to the identification of the factors regulating the growing cycle of algae very common in mountain running waters (*Hydrurus foetidus*) and to lead to the identification of secondary metabolites of particular interest for chemotaxonomical or ecological (study of the role of the compound in nature) reasons. In particular, special attention will be devoted to the possible use of scytonemin, a photoptotective pigment typical of cyanobacterial taxa which are periodically exposed in conditions of reduced metabolism, as indicator of the ultraviolet environment and of hydrological stability. In this context also the possibility of using the presence/absence of water mites species exclusive to springs as an indication of hydrological stability/instability, reflecting the intensity of the impact of climatic changes on the territory, will be evaluated with special care.

IV: *Long-term Research, Global Change and Seasonal Features*

Long-term ecological researches, started at the beginning of the 1990s in six mountain springs located in the Adamello-Brenta Natural Park, are further developed and improved and changes in physical, chemical and biological characteristics are used as environmental change indicators (climate change, increasing of N deposition, acid/base and trophic status variations, and increasing of UV radiation). This part will need the assessment of the seasonal fluctuations of the physical, chemical and biological characteristics, for a more effective separation of the long-term changes from the seasonal fluctuations. Every possible effort will be done for linking this initiative in the Southern Alps to a similar research which is being carried out north of the Alps, in the Berchtesgaden National Park (Bavaria), by Dr. Reinhard Gerecke. First attempts will be done to apply paleolimnological techniques to sediment cores taken in limnocrene springs with sufficient sedimentation rates and undisturbed sediment accumulation zones for reconstructing recent environmental changes (last 200 years). Explorative analyses (mineralogy, isotopes etc.) on different kind of carbonate concretions, present in several spring types, will be performed with the aim to explore their potentialities as environmental archives.

The six springs investigated since the beginning of the 1990s are being considered for inclusion in the Italian net of LTER (long term ecological research) sites (site name: "Sorgenti alpine trentine"; http://www3.corpoforestale.it/

V: *Development of Implemented Procedures and Criteria for Spring Integrity/Naturalness Evaluation*

The large amount of data collected will be reconsidered in order to develop implemented procedures and criteria for the evaluation of springs' integrity/naturalness. One goal will be the adaptation to springs of the Alps of an index developed in Germany (Fischer, 1996) and based on the strength of the link of the zoobenthos taxa with the spring environment (in alternative a new index could be developed). Diatom-based indices for the assessment of trophic (e.g. Rott et al., 1999) and acid/base status (Coring, 1993) evaluation of running waters will be applied to the springs studied and, if necessary, adapted to these special environments.

VI: Educational Activities to Foster Conservation

Acquired knowledge will be popularised thanks to the planning and carrying out of several initiatives (Internet site, popular book, nature-discovery path), being aware that these environment will be preserved only if their relevance will be understood by a wider audience. The recently established Internet site on the springs of the Alps (www.alpenquellen.com) will be a strategic tool for this part of the project.

The Stenico Karstic Springs

The main morphological, physical and chemical features are summarised in Table 1. From an ecomorphological point of view all the springs considered are flowing springs (rheocrenes). From a hydrogeological point of view they are karstic springs. The main lithology of the springs and of the drainage basins is limestones, with small amounts of dolomite and, occasionally, intercalated chert layers (Venzo, 1957). Rio Bianco and Rio Cugol are large sources emerging from openings (almost small caves, in particular in the case of Rio Cugol) between carbonate rock layers on steep slopes. These environments are very shaded and bryophytes are abundant (Table 1).

The Malea springs (Grande = Large and Piccola = Small) are rather different: These are small rheocrene springs emerging on a meadow surrounded by beech woods. Light availability is much higher and bryophytes less abundant. The main algae and cyanobacteria forming macroscopic structures observed during the sampling (Table 1) were the chrysophyte (golden alga) *Hydrurus foetidus*, the xanthophytes (yellow algae) *Vaucheria* sp. and *Tribonema* sp. (both in the Rio Bianco and in the Malea Grande) and the cyanobacteria *Chamaesiphon polonicus* (Rostaf.) Hansg. and *Phormidium subfuscum* (Kütz.) ex Gom. (Malea Grande spring).

Table 1: Main morphological, physical and chemical parameters and distribution of some algae and cyanobacteria forming macroscopic structures

		Springs			
		Rio Bianco	*Rio Cugol*	*Malea G.*	*Matea P.*
Longitude E		10° 50′ 51″	10° 50′ 59″	10° 51′ 12″	10° 51′ 12″
Latitude N		46° 03′ 19″	46° 03′ 31″	46° 03′ 27″	46° 03′ 27″
Altitude	m a.s.l.	668	700	710	705
Shading	%	75	95	0	40
Briophyte cover	%	85	85	80	40
Max current vel.	cm s^{-1}	136	104	176	91
Discharge	L s^{-1}	200-250	100-125	25-30	18
Temperature	°C	6.2	5.8	5.9	5.9
pH		7.78	7.77	7.70	-
Conductivity 20°C	µS cm^{-1}	146	132	134	-
Alkalinity	µeq L^{-1}	1660	1500	1500	-
$N\text{-}NO_3$	µg L^{-1}	425	481	353	-
$N\text{-}NH_4$	µg L^{-1}	11	28	104	-
TN	µg L^{-1}	482	561	472	-
SRP	µg L^{-1}	4	4	5	-
TP	µg L^{-1}	5	5	5	-
SO_4	mg L^{-1}	1.23	1.16	1.16	-
SiO_2	mg L^{-1}	1.5	0.9	0.9	-
CHRYSOPHYTE					
Hydrurus foetidus		r		++	+
XANTHOPHYTES					
Vaucheria sp. (+ *Tribonema sp.*)		+		+	
CYANOBACTERIA					
Chamaesiphon polonicus				+	
Phormidium spp.				+	+

(r = rare, + = frequent, ++ = very abundant).

From a chemical point of view (cf. conductivity and pH), the waters of the Rio Cugol and Malea Grande springs are very similar. The waters of the Rio Bianco have a slightly higher mineral content, with higher calcium and bicarbonate values (not shown in Table 1), but also with higher silica concentration. Phosphorus values are very low and coherent with a condition of oligotrophy. Nitrate values are low and comparable to those found in near-natural springs of the area (Cantonati, 1998a). Ammonia values are low, with the exception of Malea Grande, which emerges in a meadow and is reached by footpath uphill.

Diatom assemblages include high proportions of taxa of the genus *Diadesmis* and of *Diatoma mesodon* in the cases of the Rio Bianco and Rio Cugol springs, and of the Malea springs respectively (Figure 2). Values of the

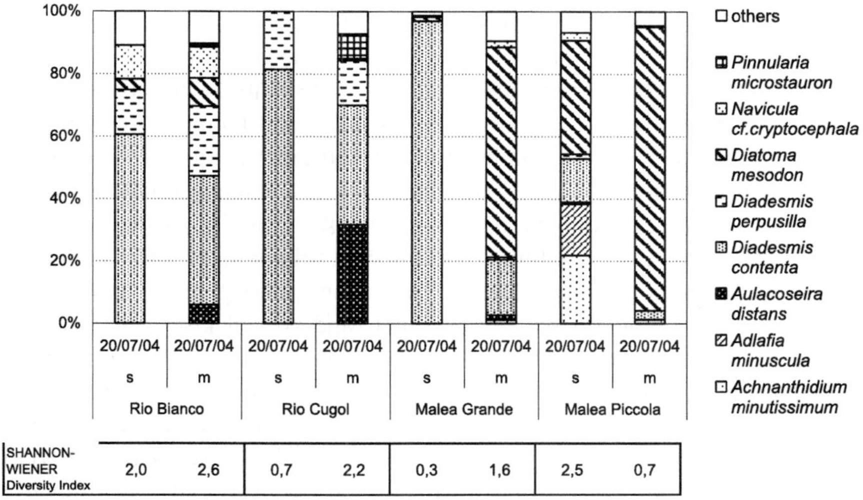

Figure 2: Composition and diversity of the diatom assemblages in the Stenico springs.

Shannon-Wiener Diversity Index are low to medium, with minimum values caused by marked dominances of individual taxa: *Diadesmis contenta* (Grunow) Mann and *Diatoma mesodon*.

Discussion

The CRENODAT Project aims mainly at a thorough documentation of the biota, morphology and physicochemistry of a representative subsample of springs in near-natural conditions of the Autonomous Province of Trento, an area of the Southern Alps characterised by high heterogeneity of lithologies and by extended altitudinal gradients. These data are intended to form the basis both for the development of integrated evaluation procedures of springs and of the documentation of the biodiversity of these habitats. This last scope is considered to be particularly coherent with the mission of the Museum of Natural History of Trento (co-ordinating institution), which is well equipped also to store the biological materials in collections. This habitat documentation is also intended to be useful for a sustainable use of the environment for tourism purposes. Therefore, also springs well known to the population and with a prominent role in the landscape, like the Rio Bianco springs (the largest and most visible of the Stenico springs), should be considered by the project.

The Stenico springs are karstic springs emerging at different altitudes that correspond to different levels of the karstic system (Venzo, 1957). The Rio Bianco springs are fed mainly by a high altitude drainage basin. This determines marked seasonal discharge fluctuations, leading to total dryness in the winter months, e.g. in the case of the Rio Bianco springs. This is clearly indicated by the diatom assemblages found that are, in the case of the Rio Bianco and Rio Cugol, dominated by well-known xerotolerant (= capable of tolerating periods of air exposure) taxa of the genus *Diadesmis*. These springs are also very shaded and comparisons with the literature— especially with Villeneuve & Pienitz (1998) that found *Diadesmis* in a shaded and temporary spring and report that this diatom was dominant also in different karstic environments, e.g. in small Turkish caves—seems to suggest that *D. contenta* and *D. perpusilla* (Grunow) Mann are not only xerotolerant, but also scyaphilic (= shade-loving) taxa. Xerotolerant taxa were found also amongst subdominant and rare diatom taxa (e.g. *Luticola mutica* (Kützing) Mann) and among the cyanobacteria (*Chamaesiphon polonicus* has strong sheets and is well known to colonise aquatic habitats that undergo desiccation). *Diatoma mesodon*, which was found to frequently dominate the Malea springs assemblages, is meanwhile well known as one of the most frequent and abundant taxa in mountain springs and streams (cf. e.g. Cantonati, 1998b). This diatom is also known (Cox, 1990) to thrive well in sunny places and this underpins the clear difference in light availability, that further differentiates the Malea springs, where *Hydrurus foetidus* was well developed, from the shaded Rio Bianco and Rio Cugol springs, where bryophytes were in part more abundant. Also the higher diversity of the diatom assemblages found on bryophytes confirms the findings of previous investigations (Cantonati, 1998b).

The taxa-poor diatom assemblages found in the Stenico springs are obviously not of great value for biodiversity assessments, but nevertheless of special interest for the very precise and strong environmental signal (very pronounced seasonal discharge fluctuations) they indicate.

Acknowledgements

We are grateful to the University and Scientific Research Service of the Autonomous Province of Trento for funding the CRENODAT Project and to Dott. Flavio Corradini (Agriculture Institute of S. Michele all'Adige – TN) for making available hydrochemical data on the springs of Stenico.

References

Cantonati, M. (Ed.), 1998. Le sorgenti del Parco Adamello-Brenta (In Italian with English and German summaries). Parco Documenti, 11. Parco Adamello-Brenta, Via Nazionale 12, I-38080 Strembo, Trento, 177 pp.

Cantonati, M., 1998b. Diatom communities of springs in the Southern Alps. *Diatom Res.*, **13(2):** 201-220.

Cantonati, M. and Ortler, K., 1998. Using spring biota of pristine mountain areas for long term monitoring. *In:* Haigh, M.J., Krecek, J., Rajwar, G.S. and Mariame, M.P. (Eds.) Hydrology, Water Resources and Ecology in Headwaters (Proceedings of the Headwater'98 Conference held at Merano/Meran, Italy, April 1998). *IAHS Publ.*, **248:** 379-385.

Cantonati, M. and Maiorana, S., 2001. Ricerche ecologiche di lungo termine in sorgenti del Parco Naturale Adamello-Brenta (Trentino). Atti dell'XI Congresso Nazionale della Società Italiana di Ecologia (SitE), Sabaudia, 12-14 Settembre 2001.

Cantonati, M., Corradini, G., Jüttner, I. and Cox, J., 2001. Diatom assemblages in high mountain streams of the Alps and the Himalaya. Proceedings of the International Symposium Algae and Extreme Environments. Trebon, Czech Republic, 11-16 September 2000. *Nova Hedwigia, Beiheft,* **123:** 37-62.

Cantonati, M., Gerecke, R. and Bertuzzi, E., 2005. Springs of the Alps, sensitive ecosystems to environmental change: From biodiversity assessments to long-term studies. *In:* Lami, A. and A. Boggero (Eds.) Developments of Hydrobiology: Ecology of High Altitude Aquatic Systems in the Alps. Hydrobiologia.

Cantonati, M., Spitale, D. and Bertuzzi, E. (Eds.) in prep. The spring habitat: biota and sampling methods. *In*: *Monografie del Museo Tridentino di Scienze Naturali*, Trento.

Coring, E., 1993. Zum indikationswert benthischer Diatomeengesellschaften in basenarmen Fließgewässern. Dissertation, Univ. Göttingen. Verlag Shaker. Reihe Biologie. ISBN 3-86111-555-7, pp. 165.

Cox, E.J., 1990. Studies on the algae of a small softwater stream. I. Occurrence and distribution with particular reference to the diatoms. *Archiv für Hydrobiologie./ Suppl.,* **83:** 525-552.

Crema, S., Ferrarese, U., Golo, D., Modena, P., Sambugar, B. and Gerecke, R., 1996. Ricerche sulla fauna bentonica ed interstiziale di ambienti sorgentizi in area alpina e prealpina. Centro di Ecologia Alpina, Trento, Report n. **8:** 1-104.

Fischer, J., 1996. Bewertungsverfahren zur Quellfauna. *Crunoecia,* **5:** 227-240.

Fritscher, J., 2004. Untersuchungen über Sulfid-Schwefelquellen in Bayern, Beiträge zum ökologischen Monitoring und zur Entwicklung von biotechnologischen Methoden für die Grundwasserreinigung. Dissertation zur Erlangung des Doktorgrades der Naturwissenschaften. Universität Erlangen, pp. 215.

Füreder, L., Vacha, C., Amprosi, K., Bühler, S., Hansen, C.M.E. and Moritz, C., 2002. Reference conditions of alpine streams: Physical habitat and ecology. *Water, Air and Soil Pollution,* **2:** 275-294.

Gerecke, R., Franz, H. and Schrankel, I., 2002. Fünf Jahre Quellforschung im Nationalpark Berchtesgaden. Mitt. Nationalpark Berchtesgaden. *Nationalpark Berchtesgaden, Forschungsbericht,* **46:** 67-72.

Haseke, H., 1998. Nationalpark o.ö. Kalkalpen (Upper Austria): Karst research Program. The Nationalpark Karst Programm 1994-1998. Homepageversion im Internet: http://ftp-waldoek.boku.ac.at/kalkalp/

Illies, J. and Botosaneanu, L., 1963. Problèmes e méthodes de la classification écologique des eaux courantes, considerées surtout du point du vue faunistique. *Mitteilungen der Internationalen Vereinigung für Theoretische und Angewandte Limnologie,* **12:** 1-57.

Kadlubowska, J.Z., 1985. Untersuchungen der Stetigkeit der Diatomeengesellschaften aus der Salzquelle, den Limnokrenen und dem Moortümpel. *Verh. Internat. Verein. Limnol.*, **22:** 2834-2837.

Odum, E.P., 1971. Fundamentals of Ecology. Saunders, Philadelphia (3rd edition), pp. 564.

Roca, J.R., 1990. Tipología físico-química de las fuentes de los Pireneos centrales: Síntesis regional. *Limnetica,* **6:** 57-78.

Rott, E., Pfister, P., Pipp, E., Pall, K., Binder, N. and Ortler, K., 1999. Projekt BMLF:

Indikationslisten für Aufwuchsalgen in Fließgewässern Österreichs, Teil 2: Trophieindikation und autökologischen Anmerkungen - WWK, Bundesministerium Land- und Forstwirtschaft, Wien. ISBN 3-85 174-25-4, pp. 248.

Round, F.E., 1981. The Ecology of Algae. Cambridge University Press, pp. 653.

Sabater, S. and Roca, J.R., 1990. Some factors affecting distribution of diatom assemblages in Pyrenean springs. *Freshwat. Biol.*, **24:** 493-507.

Sabater, S. and Roca, J.R., 1992. Ecological and biogeographical aspects of diatom distribution in Pyrenean springs. *Br. Phycol. J.*, **27:** 203-213.

Venzo, G., 1957. Le sorgenti di Stenico nelle Giudicarie (Trentino occidentale). *Studi Trentini di Scienze Naturali,* **34(3):** 3-12.

Villeneuve, V. and Pienitz, R., 1998. Composition diatomifère de quatre sources thermales au Canada, en Islande et au Japon. *Diatom Research,* **13(1):** 149-175.

Ward, J.V., 1994. Ecology of alpine streams. *Freshwater Biology,* **32:** 277-294.

Weigand, E., 1998. Limnologisch-faunistiche Characterisierung von Karstquellen, Quellbächen und unterirdischen Gewässern nach Choriotopen und biozönotischen Gewässerregionen (Nationalpark o.ö. Kalkalpen, Österreich). Unpublished Research, Nationalpark o.ö. Kalkalpen, pp. 173.

Werum, M., 2001. Die Kieselalgengesellschaften in Quellen. Abhängigkeit von Geologie und anthropogener Beeinflussung in Hessen (Bundesrepublik Deutschland) – Hessisches Landesamt für Umwelt und Geologie, pp. 273. ISBN 3-89026-331-3.

Werum, M. and Lange-Bertalot, H., 2004. Diatoms in springs from Central Europe and elsewhere under the influence of hydrogeology and anthropogenic impacts. *Iconographia Diatomologica,* 13. A. R. G. Gantner Verlag K. G., Ruggell, pp. 480.

Zollhöfer, J.M., 1997. Quellen die unbekannten Biotope: erfassen, bewerten, schützen. *Bristol-Schriftenreihe,* **6:** 1-153, Bristol-Stiftung, Zürich.

Zollhöfer, J., Brunke, M. and Gonser, T., 2000. A spring typology integrating habitat variables and fauna. *Archiv für Hydrobiologie, Supplemente 121/3-4, Monographical Studies* 349-376.

SECTION C

Quality, Pollution and Management of Water Resources

12

Water as a Symbol of National Identity in Norway

Helena Nynas
Norwegian Water Resources and Energy Directorate
PB 5091 Maj.
N-0301 Oslo, Norway

Introduction

"Nature" is one of the culturally most loaded concepts. It is one of the central concepts for human understanding and structuring of the surroundings (Glacken, 1967). The concept of nature is deeply embedded in our understanding of the world, so deep that we seldom reflect its central role in our thinking. With impulses from symbol theory (Elias, 1991), environmental history (Berntsen, 1994; Nash, 1982) and social sciences (Berger and Luckmann, 1987), this article examines nature, not as a physical reality, but as a symbol of ideas, norms and values in society. "Nature" is understood and interpreted in a historical, social and cultural context. Views of nature are:

(a) closely related to a society's prevailing norms, values and power-relations;
(b) paradigmatic—deeply embedded conceptions within a society; and
(c) closely related to use of nature both by individuals and collectively.

This article deals with culturally conditioned understanding of nature, more specifically with Norwegian conceptions of "water" as a powerful symbol in the political project of building and modernizing the nation. Water and the waterfalls were, and still are, crucial symbolic elements in building the nation of Norway and upholding Norwegian identity. At the same time, water is the quintessential symbol of the modernization and industrialization of 20th century Norway.

This article makes conscious generalization and short-cuts on the topic "humans – nature relationship". The starting point is nature as a relative phenomenon. It emphasizes nature as mirroring human beings in their historical and socio-cultural contexts.

Nature, Nation and Civilization—Etymological Perspectives

The words nature and nation have the same etymological background—the Latin word "*natio*" which means birth. The word "*nat*" relates to the innate or inherent. It refers implicitly to the origin of a nation. Both nature and nation are related not only etymologically, but also in the sense that the political and symbolic project of building a nation makes use of particular pieces of its natural basis and uses these particular pieces to distinguish its own *character*.

The word nature has at least 10 different meanings (Webster's Dictionary). A glimpse at three of them will put the concept into perspective:

(a) the natural world as it exists without human beings or civilization;
(b) the particular combination of qualities belonging to a person, animal, thing or class by birth, origin or constitution, native or inherent character;
(c) the laws and principles that guide the universe or an individual.

A culturally conditioned nature-concept challenges the conventional meanings, as it suggests nature to be understood indistinguishable from the human beings' conceptualization of the so-called "outer world".

The word civilization has at least seven different meanings. It is usually thought of as "... an advanced state of human society in which a high level of cultural science and government has been reached (…) modern comforts and conveniences, as made possible by science and technology". This meaning of the word is quite outdated today. It evaluates society and culture from a typical, and restricted, European-American point of view. Today the word civilization is mostly referred to "any type of culture, society etc. of a specific place, time or group", as for example Mesopotamian or Greek civilization.

Norway—Blessed with Water

Norway is a young nation in many respects. The whole region was covered with ice only 10,000 years ago, and in 2005 Norway celebrated 100 years of independence as a modern nation. Norway is also full of water—it has 240,000 lakes and ponds larger than 0.25 ha, and about 4000 rivers draining to the sea. Eighteen lakes are larger than 50 km^2. Some of the waterfalls are among the highest in the world. The highest waterfall has a vertical fall of 300 m, and the number of waterfalls with more than a 10 m fall is 113. The mean annual runoff is 1140 mm. The topography is steep along the west coast, and slopes gently towards the south-east. The country is also blessed with the warm Gulf current, making living conditions quite favourable compared to other countries at similar latitudes. The water resources in Norway are also almost totally under national control and governance. Only a couple of rivers in the north and east are shared with the neighbouring countries Russia, Finland and Sweden.

Together these conditions make Norway a country of abundant water. The volumes are well distributed all over the country. Droughts are a rarity. Not only the volume, but also distribution, natural storage and high heads provide excellent conditions for using water in different production processes, hydropower in particular.

From the Middle Ages until 1900 it was common that small groups of farmers collaborated in the building of small water wheels for producing flour. This meant that the farmers were locally self-contained for the energy required to produce food. It was not necessary to establish central governance for one or two main rivers—as known in other parts of the world, e.g. the Nile in Egypt. This paved the way for what can be called a "democratic" character of water in Norway (Tvedt, 1997). Even before AD 1350 rivers were used for floating timber. During the period AD 1500-1900, timber was the most important export from Norway. From early AD 1900 the hydroelectric power industry became the most important factor in the modernization of Norway.

Norwegian Nature—Water as a Romantic Topos in the Representation of Norway

Norway celebrates 100 years as an independent nation in 2005. In 1905, as Norway broke the union with Sweden, Norwegian artists had for decades contributed to national romanticism with works in which water is a crucial ingredient. A Norwegian national romantic painting is almost unthinkable without water as an element, preferably flowing or cascading and always in abundance. In this manner painters contributed in the shaping of images of the nation's nature, its essential *character* and thus to the political project of building the nation.

A waterfall, intact, pure and forever falling from the mountains, is in this perspective a *cultural* as much as a natural heritage. Conservationists at that time argued for the cultural as well as the scientific importance of taking care of nature. This reflects that the intention at that time was to create and uphold clearly distinguishable images of people in a specific territory.

Early tourism in Norway (1800s) was based on the fascination for the wild nature of the country. Norwegian waterfalls were presented as attractions for tourists as early as the 1800s. Arranging salmon fishing for foreign lords from polluted central Europe was also common. Grand tourist hotels were built to host foreigners that came to admire the fjords, mountains, glaciers and cascading waterfalls. Even today the waters in Norway are heavily represented in promoting the country as an attractive destination.

Hiking tours in the woods or mountains, as opposed to urban walks, are clearly *the* distinguished way of using leisure time for Norwegians, and is *the* preferred activity. This also means that Norwegians themselves are easy to mobilize when it comes to taking care of nature generally. It affects their daily lives.

A Nation's Park—Shaping a Character

Building and upholding the nation is a symbolic and political project. Various elements, both natural and cultural, are used to strengthen particular characteristics of the people and a region. In Norway, as in other nations, particular "parts" of nature have been specially taken care of through nature conservation. Any self respecting nation has protected and conserved parts of "its own" nature, as well as culture. Establishing a national park can be interpreted as a nation's maintenance of an image emphasizing *"pure nature"*, as it was thought to appear before civilization and the subsequent exploitation of it.

The concept of the earliest national parks is based on an idea of a national heritage, and on the wish to document the nation's natural history. The movement started in the USA in the early 1800s. The arguments for preserving specific areas were heavily associated with shaping the *character* of America in relation to the rest of the world:

> "...what a beautiful and thrilling specimen for America to preserve and hold up to the view of her refined citizens and the world, in future ages! A *nation's Park*, containing man and beast, in all its wild(ness) and freshness of their nature's beauty (...) the government would protect them in a magnificent park."[1]

The earliest national parks and preservation of elements in nature represent strong symbols of a nation's understanding of self. They symbolize the nation's "own" character and authenticity. Arguments, such as the need for scientific reference areas and the maintenance of biological diversity, have been used since the 1970s. These arguments are the most accepted ones today.

The actual establishment of national parks came relatively late in Norway. The first national park was the mountain range Rondane in 1962. In Norway the national parks (30 today) are first and foremost mountain areas. But the ideas of and suggestions for establishing parks and monuments were on the agenda before 1900. The need for preserving *actual* nature was argued for, as the painters' and writers' descriptions of nature were viewed as far from sufficient. With reference to national parks in America, Professor Yngvar Nielsen, head of the Norwegian Tourist Association, suggested in a speech in 1904:

> "Should there not be reason to think of something similar here. To "reserve" a piece of Norway. To let it keep its original nature for coming generations. There are several areas that could be suited for such use. We have the Jotunheimen. It gives itself naturally to the thought that it

---

[1] Articulated by the American painter and preservationist G. Catlin in the 1830s, cited in Nash, op. cit., p. 101.

> should be as it is, like a home for true nature, a picture of Norway, as it is today (…) What wouldn't such a natural reserve offer allurements for coming generations!"[2]

To protect the "pure and untouched nature" was viewed as important for the people and its cultural development—at that time often in terms of "civilized citizens". Hence its importance was not only for nature itself and scientific interest. Taking care of nature was meant to have a therapeutic function for the people, often as an answer to the dark sides of urbanization. It was also argued that the Nordic countries' nature and national parks would be a counterweight to southern and central Europe's zoos and gardens in the crowded cities—often characterized as "animal prisons". As the Nordic countries are sparsely populated and far from crowded, it was not too problematic to propose protection of nature. This early form of nature protection was culturally justified. The spokesmen were from academia—geographers, zoologists, botanists etc. They had scientific motives, but above all it was a favourable time to propose protection with national romantic overtones current in the young nation Norway. It was important to create and uphold an image of Norway on the same level as the nations it preferred to compare itself with.

Protection of Water as an Important Natural Feature in Norway

Protection of so called natural monuments[3] came quite early in Norway. The waterfall Vettisfossen (275 m vertical fall) was given this status, and protected in 1924. The fact that a waterfall as such was, and is attached great importance, is not incidental. The waterfall is supposed to symbolize the nation's *particular nature*. It is also the waterfall that has contributed strongly to the *development of the modern nation* Norway. It is in fact the symbol of identity (the past) and progress (the future) in Norway. The entire project of modernization of society in Norway is based on the use of the energy of falling water to produce hydroelectric power.

The development of hydropower put considerable pressure on the watercourses in Norway. Around 1960 the debate started on exempting rivers from development. The Norwegian environmental movement is strongly based in the struggle for protecting waterfalls and rivers against hydropower production.

It is also remarkable that the bird fossekallen (in Latin *Cinclus cinclus*) was elected by the people as the national bird in Norway in the 1960s. It is a tiny bird living *near water, and off insects, particularly in rapids*. Here a

---

[2] Speech printed in: Aanesen, op. cit., pp. 209-210 (translated by author).

[3] "Monument" is here illustrating—it signifies that nature protection was of cultural importance.

water-dependent little animal gets national status—a struggle with the natural forces in rivers! Even a Norwegian "national waterfall" has been the topic of a popular vote. This illustrates the importance and status of water in Norwegian minds.

Water as the Key Factor in Modernizing Norway

Norway is known as the land of water. The topography is in many places very suitable for using the waterfalls for hydroelectric purposes. The development of hydropower started in 1880s and expanded rapidly after the Second World War. After the war Norway embarked on a modernizing programme. The waterfalls were called "the white coal" as opposed to black and polluting coal in central Europe. The modernization of households and industry led to an increasing demand for electricity—in Norway hydropower development. Today 99% of all electricity produced has its source in hydropower. Norway is the sixth largest hydropower producer in the world and the second largest in Europe.

One illustrative example on the relations between traditional and modern times in Norway was painted, and literally pointed out, by Theodor Kittelsen in 1908 (Christensen, 2002). Kittelsen was fascinated by the advances of modern times, and received an order for paintings of Rjukanfossen, a waterfall at that time under hydropower development. The man who commissioned the paintings was Sam Eyde, a big hydropower-entrepreneur at that time. The Kittelsen's paintings give many a view on the traditionalists struggling against modern times and foreign capital *vs* the entrepreneurs and the belief in industrialization as based on the waterfall. The paintings illustrate the traditional Norway "losing its virginity" *and* the strong progressive Norway "blessed with hydropower-possibilities". This can also be seen as an illustration of the relations between art and capital. Artists were not only traditionalists, but acted also in favour of modernity.

Hydropower and Water Protection Today

Today 389 rivers are permanently protected from hydropower development. Four of the 12 highest waterfalls are protected. Of the hydropower potential (total 188 TWh) some 118 TWh is developed. About 23% of the potential is permanently protected. The hydropower developers are today meeting a political climate of "enough is enough", and only minor developments and rehabilitation of existing power-stations is licensed. The consideration of protection interests is an essential part of a normal licensing procedure.

Since the amounts of water are huge in Norway, it has not been necessary to develop alternative sources of energy on a large scale. Nonetheless, growing demands on energy still apply pressure to further developments in this field.

Norwegian waters have given Norway a high-tech status and a modern standard of living worth envying, but they also have—in their amounts and dominance as energy source—blocked alternative thinking in the energy field.

The fact that almost a fourth of the Norwegian hydropower potential is still protected may be seen as a relatively great success for nature conservation, taken into account the considerable pressure for hydropower development. The strong national status water is given, and the Norwegian way of feeling and using nature on a daily basis has indeed contributed to this situation.

Conclusion

Interpreting water and the waterfall in the context of national identity, shows that ideas, norms and values are not static, but actively shaped in, and by society. They are altered over time along with other developments in society, such as technological and scientific developments. The point is also to contribute to a more nuanced view of nature. This view is a *supplement* to the more usual technical and pragmatic view of nature.

The cultural and symbolic perspective opens for reflection on, and questioning of the meanings and values we attach to intact nature/nature today. The preservation of nature can be said to contribute to a view of nature as something "out there", at a distance from daily practice. To approach the theme "nature" will require the consideration of the active shaping of particular natures in and by society.

The following reflections may serve as conclusion:

- The time of building the nation is over – the focus has shifted to either local or global/international levels. Water is running, and has always been running, without man-made boundaries. For water management it is crucial to extend the view to the river basin as such. This requires extended cooperation at all levels.
- The arguments for protecting nature (here especially water) should not underestimate the crucial meaning it has for the general well-being of humans in any society. The existence of so called untouched nature facilitates actual experience and use of this nature, and this must be seen as a necessary condition for developing an attitude of taking care of nature, both individually and collectively.
- Protecting nature in parks and reserves is set out in governmental laws and plans. Nature protected in this way is often to be experienced in built-up-centres – is this "wrapping of nature" appropriate? Yes, but it has limitations. A protectionist attitude starts with an individual's development from child to adult. Learning how nature acts and reacts, how it forms our conditions *and* how society has formed the nature historically is fundamental. Learning about nature is today dominated by a technical and natural-scientific approach. These approaches should be complemented by cultural, historical and ethical perspectives.

- Development of civilization is connected to advances in sciences. An appropriate question is whether scientific and technical advances in our modern society also contribute to alienating people from having immediate contact with, and understanding of, this society's relationship to nature.
- Social and cultural scientists have for too long neglected the fundamental questions on nature as shaped in the mind of people and as constructed by social practices. In the Nordic countries cultural geographers have made interesting contributions in this respect (Olwig, 1987; Lehtinen, 1991; Mels, 2002).
- Challenges for social and cultural scientist include:
  - Nature viewed *separately* from (as existing independently of) human beings makes no sense—our conceptualization of it is *qua* conceptualization relating to social structures (language being one of them);
  - To analyze fundamentally the *social* construction of nature (Vogel, 1996);
  - Local natures—local realities and everyday spheres of living should be made primary issues in welfare definitions (Lehtinen, op. cit., p. 149);
  - Are social and cultural scientists better off with the term *environment* than nature? Environment would include nature, people, mental, social and political structures. This means rejecting dualism, the natural world and the social world as *not* distinguishable, "because the *Umwelt*, the surrounding world of "nature", is itself in various senses the product of social practices" (Vogel, op. cit., p. 7).

References

Aanesen, E. (Ed.), 1904. *Den Norske Turistforeningens Aarbog for 1904*. Oslo, Norge.

Berger, P.L. and Luckmann, T., 1987. *Den samfundsskapte virkelighed. En videnssociologisk afhandling*. Lindhardt og Ringhof, København, Danmark.

Berntsen, B., 1994. *Grønne Linjer. Natur- Og Miljøvernets Historie i Norge*. Grøndahl og Dreyers forlag, Oslo, Norge.

Christensen, A.L., 2002. *Det Norske Landskapet*. Pax forlag, Oslo, Norge.

Elias, Norbert, 1991. The Symbol Theory. Sage, London, UK.

Glacken, C.J., 1967. Traces on the Rhodian Shore: Nature and Culture in Western Thought from Ancient Times to the End of the Eighteenth Century. University of California Press, Berkeley, USA.

Lehtinen, A.A., 1991. Northern Natures. A study on the forest question emerging within the timber-line conflict in Finland. *Fennia,* **169(1):** 57-169, Geographical Society of Finland.

Mels, T., 2002. Nature, home and scenery: the official spatialities of Swedish national parks. *Environment and Planning D: Society and Space,* **20:** 135-154, Pion, London, UK.

Nash, R., 1982. Wilderness and the American Mind. Yale University Press, New Haven, USA.

Olwig, K., 1987. *Hedens Natur: Om Natursyn og Naturanvendelse GennemTtiderne*. Teknisk forlag, København, Danmark.

Tvedt, T., 1997. *En Reise i Vannets Historie*. Cappelens forlag, Oslo, Norge.

Vogel, S., 1996. Against Nature – the Concept of Nature in Critical Theory. State University of New York Press, USA.

13

Assessing Renewable Water Resources and Water Use in Angola

David A. Wright, Kjetil Sandsbråten[1] and Olav Osvoll[2]
NVE, International Section Drammensveien 211, P.O. Box 5091
Majorstua No-0301, Oslo, Norway
[1]Sweco Grøner, Lysaker, Norway
[2]Norwegian Directorate of Water Resources and Energy
Oslo, Norway

Background

The potential of renewable water resources of Angola is not known with sufficient accuracy, mostly due to lack of data. Almost all of the 189 hydrometric stations in operation at independence in 1975 were abandoned during the years of civil war. The same is the case for data on total water use, water consumption, and future water demand. Proper water development planning is thus almost impossible. The "Rapid Water Resources and Water Use Assessment for Angola" was designed to start the ball rolling in the direction of establishing a sufficiently accurate assessment of the renewable water resources and water use that can be further elaborated and updated as data become available. As such it forms one of the steps towards achieving the goals of the national strategy for the Development of the Water Sector in Angola. The project is Activity C of the National Water Sector Management Project (NAWASMA), being carried out as institutional co-operation between the Angolan National Directorate of Water (DNA-Direcção Nacional de Águas) and the Norwegian Water Resources and Energy Directorate (NVE), under the financing of NORAD and covering the entire country. SWECO Grøner carried out the assessment for and in collaboration with DNA under the Ministry of Energy and Water. The development objective of the NAWASMA project is improved water sector management in Angola through a strengthened institutional capacity of DNA. The specific development objective of Activity C was to rapidly assess the water resources, water use, and future water demand in Angola, which would in turn help Angola to be in a position to better deal with its international relations as a headwater country with shared water resources. In addition to providing up-dated estimates of water resources and water use, the future capacity for managing and planning water resources was improved by transfer of knowledge and capability to DNA. The project also enabled DNA and the other Angolan authorities in the water sector to move on in their work with national master plans for the water sector, identifying bottlenecks for further planning and management in terms of data quality and availability.

Assessment Methodology

Relevant authorities and stakeholders in the water resources sector were involved in the study in order to include their knowledge and data in the assessment. DNA was the facilitator in this respect, providing the link between the Consultant and the various authorities. The challenges to obtaining reliable data on water resources and water use in Angola are many. In some sectors no data is freely available. The importance of identifying and addressing these bottlenecks in the development of assessments for the sector cannot be overemphasised and was a major activity during the assessment.

Another important point was the aerial resolution to be used in the study. The established boundaries of regions of Angola to be identified and used were adopted.

These are the major river basins, some of which are the headwaters of the international rivers, and the coastal rivers draining to the Atlantic Ocean (Figure 1):

- Zaire (Congo)
- Zambezi
- Okavango
- Etosha
- Coastal rivers draining to the Atlantic (including the Luando-Kwanza (Cuanza) system)

These were further disaggregated into individual catchments for both the water resources and the water use assessment totalling 77 catchments in all.

Estimation of Angola's Renewable Water Resources

For surface water mean monthly data from the hydrology records of the NAWASMA, database in DNA were transferred, assessed and used, supplemented by regional data. Monthly and yearly means were used to generate a simplified runoff map for those parts of Angola covered within the hydrometeorological network. This data was subjected to further checks before comparison with regional runoff and rainfall distribution, especially with a view to compensation for the lack of data from the eastern parts of Angola.

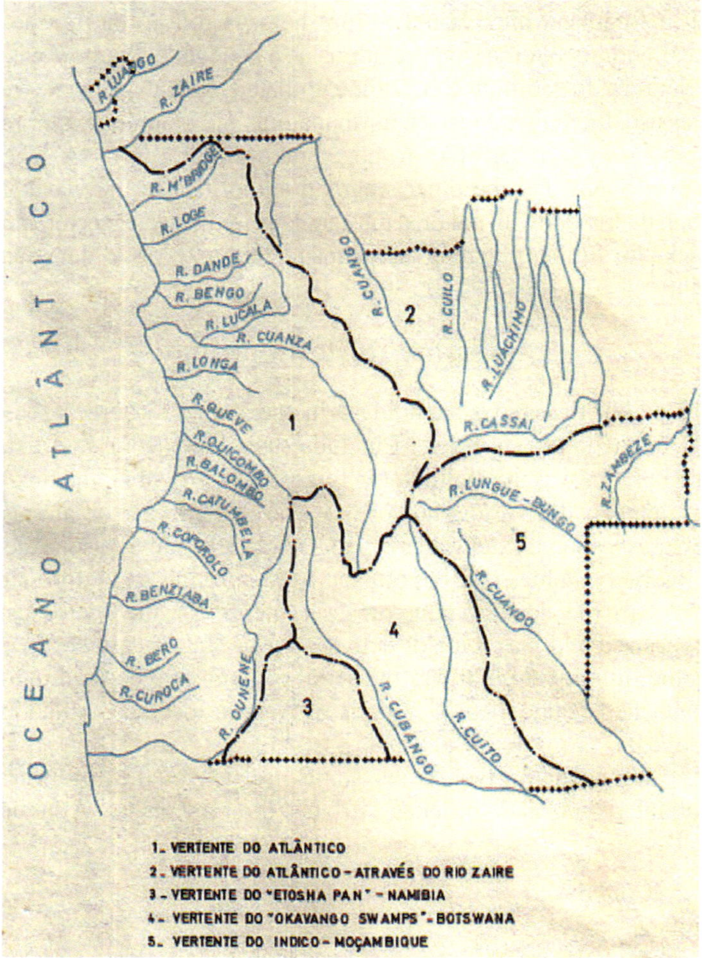

Figure 1: Major drainage areas of Angola.

INAMET, the institution responsible for meteorological records, provided data for a selection of meteorological stations. After quality control the data was compared with regional figures for rainfall distribution. The data sources for the evaluation of the hydrogeology and groundwater resources of Angola were mainly a hydrogeological map of Angola at scale 1:1,500,000 and the geological map of Angola at scale 1:1,000,000. In addition geological and hydrogeological maps and interpretations produced as the United Kingdom contribution to the International Hydrological Programme (IHP) of UNESCO Southern Africa, and the FRIEND Phase II project were accessed. Meetings were held with HYDROMINAS, UNICEF and Direcção Provincial de Águas in Lubango, and information on groundwater registers and some groundwater data was collected. These data sources gave the possibility to indicate probable groundwater yields in most of the common aquifers, but data for estimation of the volumes of renewable groundwater resources were not accessible.

Sediment Transport and Identification of Basins with Severe Soil Erosion

Data on sediment transport is virtually non-existent in Angola. Consequently, this is one of the major bottlenecks, which needs to be prioritised in DNA's further work on water resources. The Consultants were successful in obtaining a copy of a study of the sedimentation of the Cambambe Dam reservoir on the important Cuanza River[1].

The study is based on bathymetric field surveys carried out and gives estimates of the sediment volume that has accumulated in the reservoir since its implementation. Although not necessarily representative for other basins in Angola, it does at least give one data point for the important Cuanza River. Contacts were also made with institutions in neighbouring countries to seek out further possible data sources, in particular for the Okavango Delta. Internet resources were extensively researched to find any further existing relevant studies.

Some limited information was found on soil erosion caused by human activity in the catchments draining to the Atlantic Ocean. Figure 2 shows areas identified with human induced soil erosion.

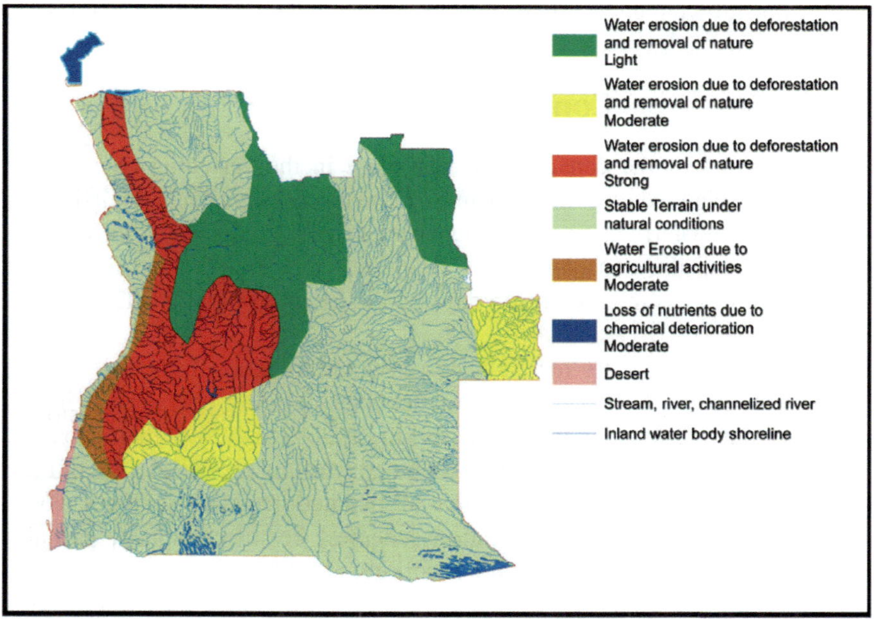

Figure 2: Human induced soil erosion in Angola.

[1] Estudo de Avaliação da Sedimentação da Albufeira da Barragem de Cambambe no Rio Cuanza, PM Consultoria Obras Hidráulicas, Luanda, February 2002

Scenarios for Urban, Peri-Urban and Rural Population Growth up to 2025

This was one of the most complex parts of the assessment due to the lack of firm population data in Angola. The methodology was developed by gathering and comparing various available sources of population data and estimates, discussing and assessing the merits and shortfalls of each one, and applying international experience combined with local knowledge. The provincial breakdowns of population estimates were even more challenging. Once established and adopted, the population estimates for each province were again disaggregated into catchment populations and entered into the GIS water resources database on a catchment basis by first splitting each province up into its various catchments and, using the satellite maps of the locations of the towns, villages and settlements in each catchment, allocating first the urban and periurban figures to the correct catchment in the province. The rural population figures were then distributed to the various catchments proportionately to the intensity of settlements in each catchment as indicated by the satellite imagery. The result is a catchment distribution of the forecasts within each province. Catchments that straddle province boundaries were dealt with as sub-catchments within each province, the total of the various provincial sub-catchments then being summed by the GIS system to arrive at total catchment figures.

Consumptive Water Use

For estimation of the consumptive water use in the agriculture sector, the existing and planned irrigation schemes, the main crops grown under irrigation and the net annual irrigation requirement were taken into consideration. Washing of animals was also considered. For irrigation activities, year-round irrigated agriculture was assumed with 365 days, while for dried periods 182 days was used (Figure 3).

The volume of water for 2005 was assumed to be as of 2004, i.e. at the time of the study. For those irrigation schemes partially operational, it was assumed that in 2005 they would be working at 20% to 40% of their capacity, whereas in 2015 these schemes were assumed working at 50% to 75% of their capacity. For 2025 it was assumed that all planned irrigation schemes would be constructed and that all irrigation schemes, be they planned, partially operational or operational, would be working at 100% of their capacity.

In relation to animal watering, projections were made based on figures provided by the Department of Animal Production of the National Directorate of Livestock, a government body under MINADER, the Ministry of Agriculture and Rural Development. As the figures provided on animals are concentrated in the southern region of Angola, a certain re-distribution of animals was made for the whole country, considering those provinces with natural conditions for livestock development.

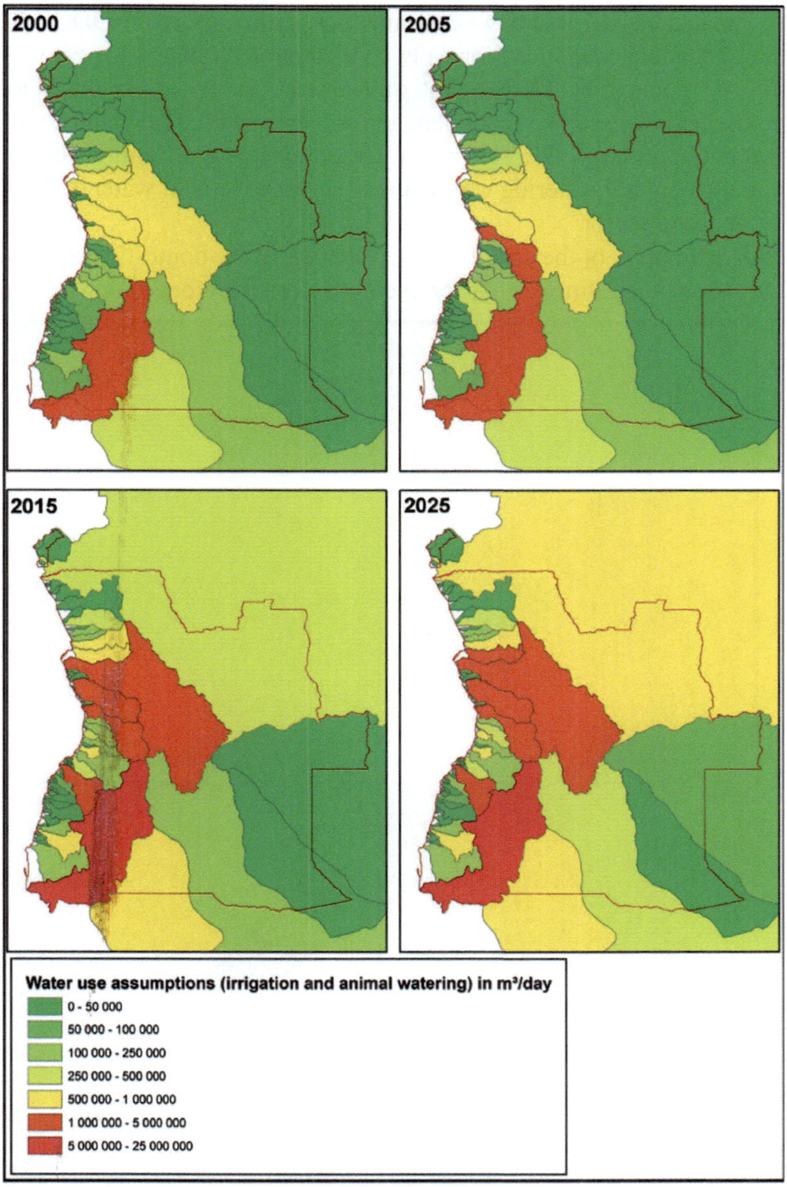

Figure 3: Catchment-wise irrigation and animal husbandry water use forecasts.

In relation to poultry development it was assumed that this activity would be revitalised in those provinces where, in the recent past, a certain level of development had been reached. The annual growth of cattle was assumed to be 3%, while the annual growth of pigs, sheep and goats was assumed to be 4%.

The same annual growth was assumed for poultry. The AGRODOK Series recommendations were used for daily consumptions of water for cattle, pigs, sheep and goats. The Department of Animal Production of the National Directorate for Livestock recommendations were used for daily consumptions of water for poultry. The sum of water volumes for irrigation and water volumes for animal watering was assumed to be the total of water demand for the agriculture sector.

For estimation of the consumptive water use in the domestic water supply sector, water use estimates and projections were developed on a per capita

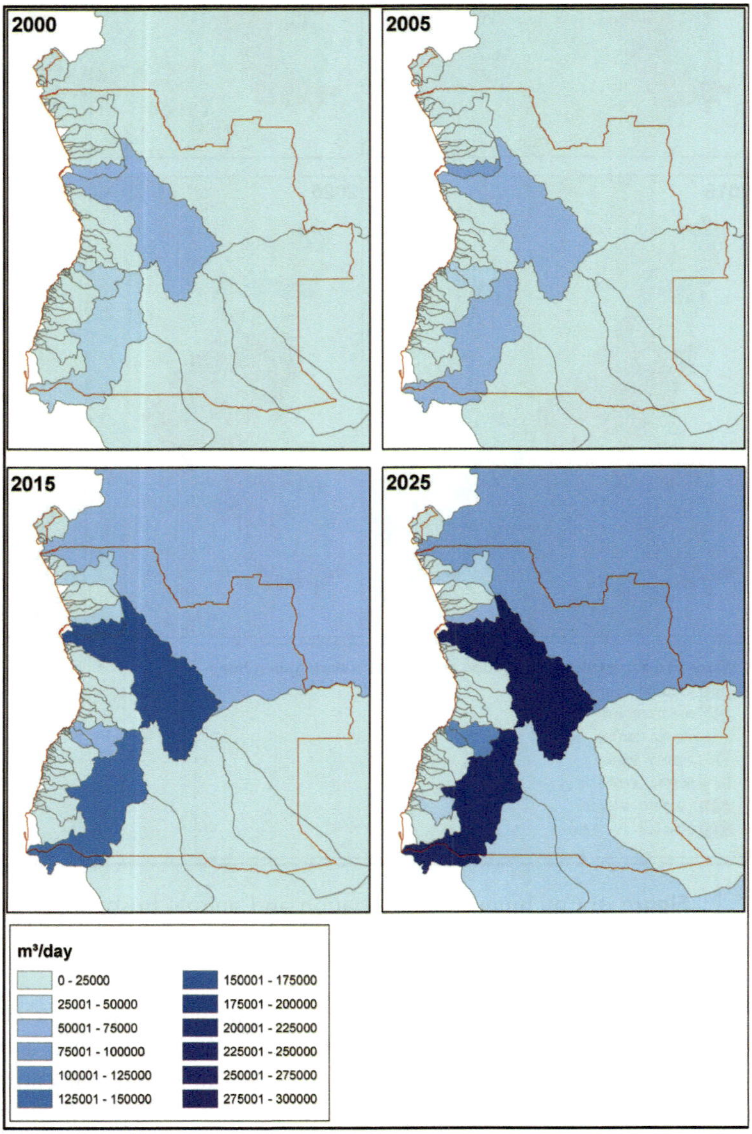

Figure 4: Catchment-wise domestic water supply forecasts.

basis and applied to the population estimates (Fig. 4). The information in the existing water master plans for towns in Angola was used in this analysis.

Water Resources/GIS Software and Database Tools Applied

Key to the establishment of criteria and the formulation of outline plans for maintaining and developing water resources and water use assessments is the system sustainability. It was realised at an early stage that the capacity of DNA in this respect is limited, as were the resources available to this project. Consequently focus was put on developing and handing over a system that could easily be used and updated for water resources, as well as a system that could be useful in DNA's other activities. The resulting Water Resources and Water User database and tools chosen are therefore based on the GIS system that can be easily adapted to almost any other use, and is well known in Angola, being used by NGOs in de-mining activities, the oil industry etc.

The available hydrometeorological, meteorological and other types of data for use as input to the system are detailed in some areas and sparse or non-existent in other areas of the country. The most appropriate and sustainable methodology for establishing the water resources and water use database with such constraints was to implement an Arc-GIS database system for the entire country. The database system is raster based and the entire country is broken down into rectangular units or rasters, their size depending on the data availability for each type of data (for example with 1 km raster for rainfall data). Each data type could then be represented by one layer of rasters and the different data layers combined using mathematical algorithms to arrive at the desired data sets for use in the water resources and water use assessment.

The advantage of this system is its flexibility in that it can easily be extended or updated as new data becomes available. In addition DNA has been provided with a powerful and useful Arc-GIS software tool which will undoubtedly also be found useful in its other activities and projects. In order to obtain maximum benefit from this system a dedicated plotter for producing large-scale printouts and visualisations from the system is advisable. Consequently the Consultant has used funds from its budget for office equipment to provide such an A1 size plotter to DNA.

Angola's International Headwaters

In a shared water resources context, Angola has several headwaters that form international rivers. It has five shared watercourses with its neighbours; namely the Zaire (Congo) River (8% in Angola), the Cunene River (87% in Angola), the Okavango River (21% in Angola), the Zambezi River (19% in Angola) and the Cuvelai/Etosha River (33% in Angola). The main characteristics of the headwaters of these international rivers are presented here.

The Cunene River Basin

The Cunene River has its headwaters in Huambo Province in the Sierra Encoco Mountains in southwestern Angola (Figure 5). The river flows in a southerly direction to the Ruacana Falls where it turns to the west and proceeds to the Atlantic Ocean. The lower section of the river cuts through a deep gorge that starts at the Ruacana Falls. In the 340 km between Ruacana and the Atlantic Ocean, the river falls more than 1100 m and this important feature provides the Cunene River Basin with a hydroelectric power potential of some 2400 MW. Between 1926 and 1969 the Portuguese and South African governments entered into three Water Use Agreements on the Cunene. In the first agreement of 1926, it was agreed that Namibia has the right to half of the flow of the Cunene, provided that a water scheme for such a purpose would be feasible.

The second agreement in 1964 related in general to the utilization of rivers of mutual interest between the parties, inferring the inclusion of other rivers like the Cuvelai and the Okavango in Angola or river systems like the Limpopo and Incomati in Mozambique as well. In that agreement the principle of best joint utilization was accepted and was defined as the allocation and utilisation, on an equitable basis, of shared water resources with a view to achieving the optimum benefit for the states concerned, within the limits of the available quantity of water. One other country, the Kingdom of Swaziland in 1967, has also acceded to this agreement.

The detailed feasibility investigations and related activities for that first phase of the development of the hydropower potential of the Cunene River

Figure 5: Cunene River basin.

and the diversion of water into northern Namibia set in motion by the 1964 agreement culminated in the Third Water Use Agreement of 1969 which initiated the construction of the proposed Cunene River Scheme. This agreement established a Permanent Joint Technical Commission (PJTC) and made provision for Namibia to abstract water at 6 m^3/s at Calueque for diversion to the Cuvelai basin in Northern Namibia. The project comprised the Gove dam to regulate the flow of the Cunene, the Calueque dam and pump station for the diversion of water into Namibia, the Ruacana weir for the diversion of water into Ruacana Power Station, and the power station itself. Of this infrastructure, the Calueque Dam was never completed due to the war in Angola at the time. The Gove Dam was completed in 1975 and the works at Ruacana in 1978. The Ruacana Power Station, with an installed capacity of 240 MW and generation capacity of 1055 GWh/year located in Namibia has not been operating at its full capacity due to the lack of continuous regulation of Cunene flows at Gove.

The total development of the Cunene River includes the multipurpose hydropower and irrigation scheme at Matala in Angola. The hydropower facilities at Matala were upgraded from 27 MW to 40 MW in 1989, but the planned land available for irrigation was not cultivated due to damage to the canal system. However, the Matala system presently has a potential of 10,000 hectares, of which 5,000 hectares were recently rehabilitated by the Brazilian contractor ODEBRECHT, and the remaining 5,000 hectares will be rehabilitated when funds become available. According to international agreement between the countries, Namibia may divert 6 m^3/s from the Cunene River at Calueque across the catchment to the Cuvelai drainage basin for supply of the domestic and irrigation water demand in northern Namibia.

In September 1990, some six months after the independence of Namibia, the governments of the Republic of Angola and Namibia endorsed and affirmed the previous agreements reached between Portugal and South Africa. The Permanent Technical Commission was reinstated and a Joint Operating Authority for the Cunene Basin was planned but has not, to the Consultant's knowledge, as yet been formalised. This bi-national authority would be created if the governments of Angola and Namibia could approve the feasibility study for the construction of either Epupa or Baynes (Marienflus) hydropower plant. Both Epupa and Baynes hydropower schemes are located in the international reach of Cunene River.

The Cuvelai River

The Cuvelai River rises in the southern foothills of the Sierra Encoco in southwestern Angola (Figure 6). It drains southwards towards the Etosha Pan in northern Namibia. The Cuvelai is perennial for about 100 km before it ramifies into a delta of ephemeral watercourses, which cross a broad plain of low relief. This delta converges again to terminate in the ephemeral Etosha

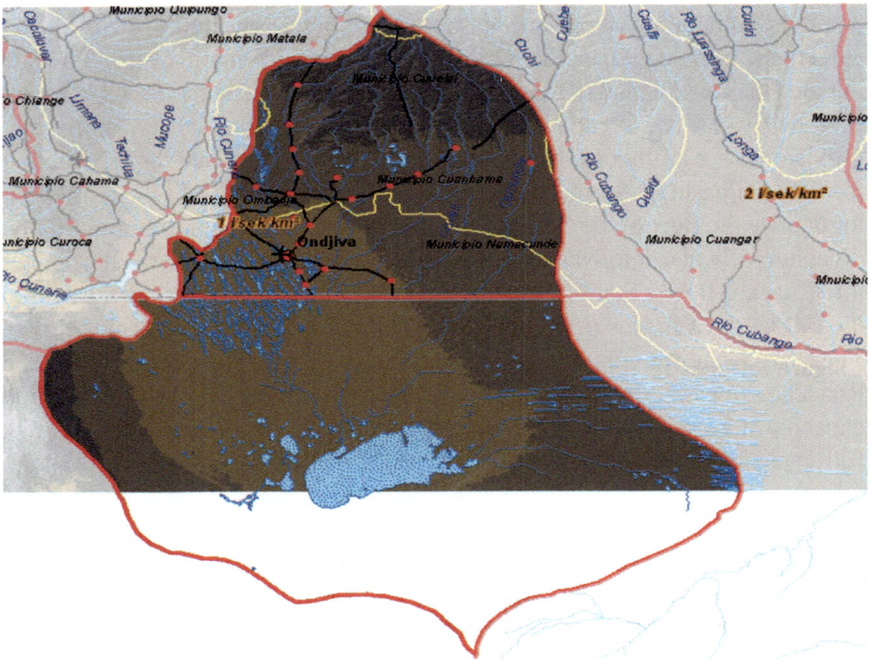

Figure 6: Cuvelai River basin.

Pan. The watercourses, called oshanas, are the lifeblood of an area where 650,000 people, or just less than half the population of Namibia live. Due to the arid climatic conditions, surface waters and shallow wells dry up from time to time. The groundwater is saline and the only way to augment these rather unreliable water supplies is to import water from the perennial Cunene River.

This is the main reason for diverting water from the Cunene River Basin to the Cuvelai Basin. The water scheme is operated by the Namibian Department of Water Affairs on Angolan territory and serves as an excellent example of cooperation between the states. The existing water supply network, distributing water through canals and pipelines to the population, is one of the largest in Southern Africa.

It is clear that any alteration to this international watercourse system in Angola or Namibia will have major repercussions for the fragile, semi-arid ecosystem and the people living on the flood plains. However, there is no specific international agreement between Angola and Namibia on water allocation or further studies in the Cuvelai Basin.

The Okavango River Basin

The Okavango River Basin rises in the southwestern Angolan highlands, near and just east of the source of the Cunene and the Cuvelai rivers (Figure 7).

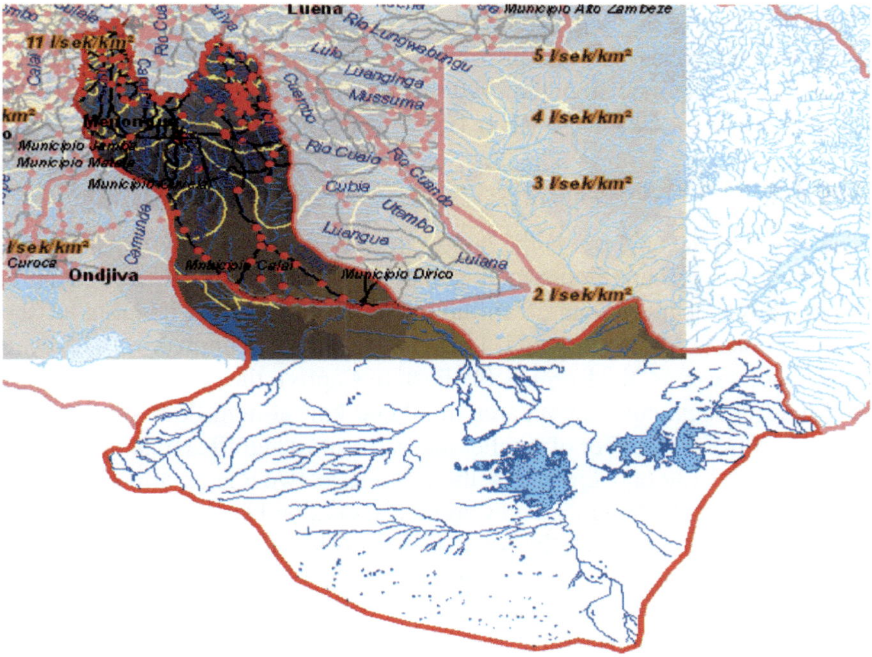

Figure 7: Okavango River basin.

The Cubango flows for more than 600 km through the upper catchment in a southerly direction until it reaches the west-east cutline border between Angola and Namibia. From that point, the river forms the border between Angola and Namibia over a distance of some 400 km. It then turns southwards again, and ends in the Okavango Swamps in Botswana. The mean annual runoff of the Okavango River at Muhembo on the border between Botswana and Namibia is 10,000 million m^3. The main tributaries of the Okavango are the perennial Cuito River and the ephemeral Omatako River. The Cuito River rises in the highlands in the central Bié Province of Angola and contributes half of the flow of the Okavango River.

It has been estimated that some 20 million m of water is abstracted per annum from the Okavango River for domestic and irrigation consumption in Namibia. A dam was built in the upper catchment of the Omatako River to divert water for domestic and industrial consumption in the Windhoek-Okahandja complex in the Swakop River catchment in central Namibia. No major development of the water resources of the Okavango River or the delta took place in Botswana, except for the Mopopi Dam. The dam was built to supply water to the Orapa diamond mine and was created by using the basin of the Putimolonwane Pan and constructing earth embankments around it to impound more water. The reservoir capacity is 100 million m^3 and it covers 24.3 km^2 at full supply level. Water is pumped into the dam from the Boteti

River, which is the outflow of the Okavango Delta. The institutional arrangements concerning the utilization of the Okavango Basin were under discussion in the first half of the 1990's that led eventually to the establishment of the Permanent Okavango River Basin Water Commission (OKACOM).

The Zambezi River Basin

The Zambezi River Basin is the largest of the African river systems flowing into the Indian Ocean (Figure 8). It is shared by eight basin states and supports a population of more than 20 million people. The major tributaries of the Zambezi rise in Angola, Malawi, Tanzania, Zambia and Zimbabwe. There are five major swamps, the Borotse, the Eastern Caprivi, the Kafue, the Busanga and the Lukanga, covering an area of 20,000 km^2 at high flood periods.

Apart from a number of smaller lakes, the most significant natural lake is Lake Malawi (30,000 km^2), but there are also two major artificial lakes, namely Kariba (5,180 km^2), and Cahora Bassa (2,660 km^2). Other reservoirs with large surface areas are the Kafue Dam (809 km^2) and the Ithezithezi Dam (365 km^2). It has been estimated that more than 160,000 tonnes of fish is caught per annum in these water bodies.

More than 28 dams with a storage capacity in excess of 12 million m^3 of which Kariba is the largest (160,000 million m^3) and Cahora Bassa the second largest (52,000 million m^3) have been built for domestic, industrial and mining, water supply, irrigation and power generation. The countries with dams are Malawi, Mozambique, Zambia and Zimbabwe. At present the major hydropower facilities are in Victoria Falls, Kafue Gorge, Kariba, Cahora Bassa, and on the Shire River at Nkula, Tedzani, Kapichira. More dams are possible at various locations.

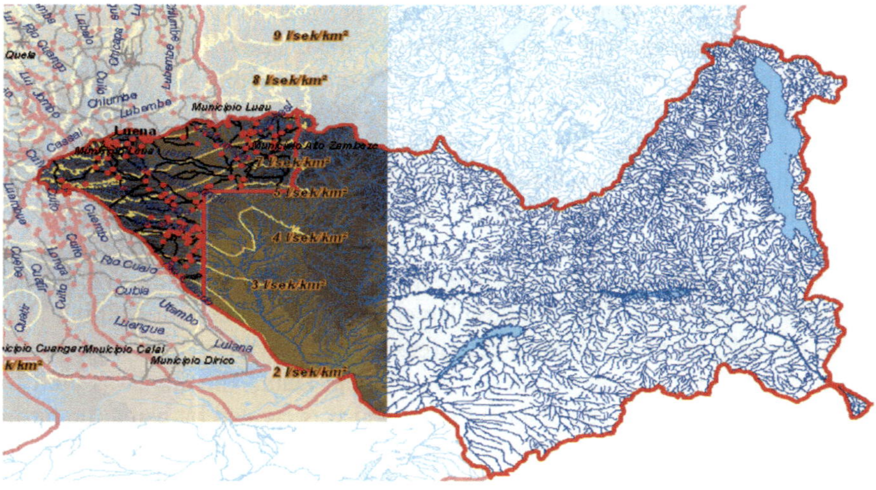

Figure 8: Zambeze River basin.

Although the available water resources in the Zambezi Basin in general exceeds the demand, this situation may deteriorate as a result of the increase in population, more industrial and mining development, increased irrigated food production, a higher standard of living of the population and by taking the environmental water demand of the system into account. However, it is envisaged that the most significant increase in water consumption will be due to irrigation projects. The Zambezi River Basin is thus clearly the main life supporting artery of eight basin states and an effective River Basin Commission is desirable to ensure its well-being. Such a commission has not as yet been established, although the Zambezi River Authority has been formed as detailed in a subsequent section.

The Zaire (Congo) River Basin

The Zaire River originates in the highlands located in eight co-basin states (Figure 9). However, most of the contribution to the runoff at the mouth of the Zaire River is generated in the middle courses of the river in the central tropical rainforests of the Zaire Basin on the equator. The flow in the upper reaches of the drainage basin is of lesser magnitude, especially in Angola, the Central African Republic, and Tanzania. The annual average runoff in the Zaire is 1,260,000 million m^3, and the average flow is 40,000 m^3/s. The historic minimum and maximum flows vary between 21,400 and 73,600 m^3/s respectively, but for 98% of the time the river flow exceeds 26,400 m^3/s.

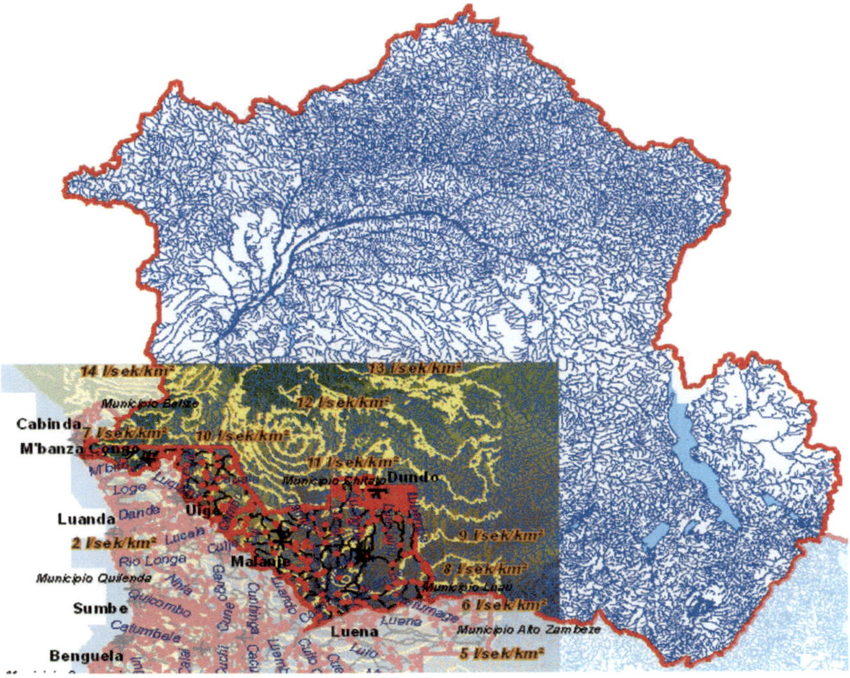

Figure 9: Congo (Zaire) River basin.

The main potential of the Congo (Zaire) River is for the generation of hydropower. There are many falls and rapids, which provide potential sites for development. The river has a total theoretical generating capacity of some 100,000 MW and only a tiny fraction of this is today developed. In spite of the many waterfalls and rapids, the Zaire River is a very important waterway because the river is navigable over long distances and provides good opportunities for boat transport and trade between the basin states. There are large wetlands and lakes in the Zaire Basin within Zambia and Tanzania,

6019	
Cuanza	
Area in km²	
150445.6	
Perimeter in km	
2701.7	
Elevation (m.a.s.l)	
Mean	Maximum
1200	1964

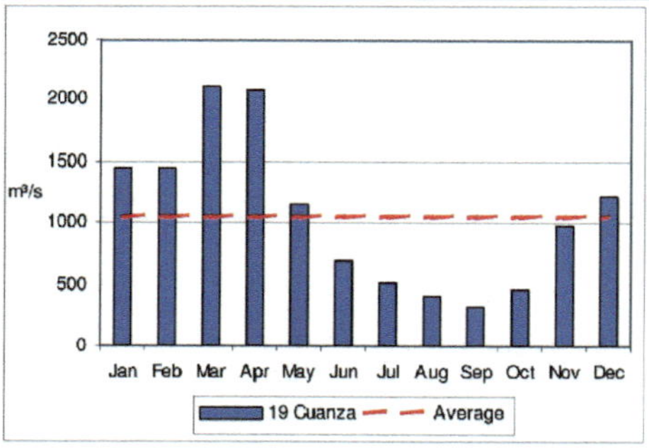

Annual mean discharge m^3/s (Q)	
1064.4	
Annual mean specific discharge (q) $l\ s^{-1}\ km^{-2}$	
Mean	*7.1*
Max	14.2
Min	1.6

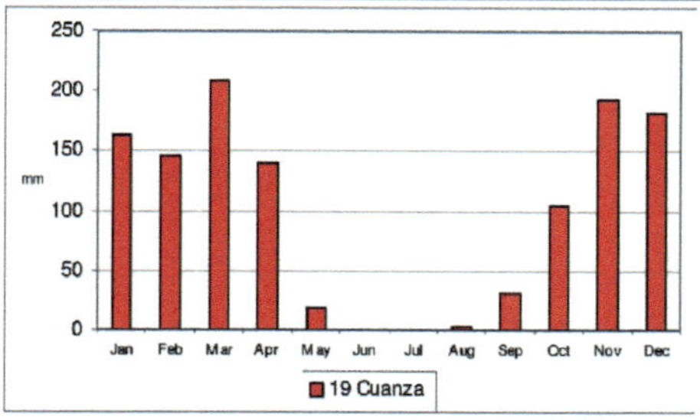

Annual precipitation in mm
1188

Population and Water use				
	2000	2005	2015	2025
Population	2844186	3411304	4622503	6221152
Water use (m³/day)	52916	67412	160486	277347
Water use (m³/s)	0.612	0.780	1.857	3.210

Animal Watering and Irrigation				
	2002	2005	2015	2025
Water use (m³/day)	509777	822747	4489858	4697133
Water use (m³/s)	5.9	9.5	52.0	54.4

Existing water deficit
NO
Future water deficit
NO

Figure 10: Example of the 77 Catchment Summary Sheets.

which provide important grazing, fish and wildlife resources to the population. About 20 large dams have been built on the tributaries of the Zaire River but none within the SADC region. Most of the dams are used for water and power supply. There are plans for further development of water or electricity supply infrastructure on the Zaire River within the SADC States.

Water Balances for the Angolan Catchments

The data gathered and incorporated in the GIS system was used to generate catchment summary sheets for each of the 77 defined Angolan catchments. An example of a catchment summary sheet is shown in Figure 10.

Stakeholder Workshop

As a conclusion to the study, a Stakeholder Workshop was held in Luanda in April 2005 in which representatives from a broad cross-section of ministries, institutions, donors and non-government organisations and other stakeholders participated. The workshop turned out to be an enormous success and resulted in the definition of the key elements of a roadmap for achieving the recommendations and goals identified in the study.

14

Water Management Issues in Middle Mountain Catchments of the Nepal Himalayas: The Downstream Perspective

Juerg Merz[1,2], Pradeep M. Dangol[2], Madhav P. Dhakal[2], Bhawani S. Dongol[2], Gopal Nakarmi[2], Pravakar B. Shah[2] and Rolf Weingartner[1]

[1]Hydrology Group, Department of Geography
University of Bern, Bern, Switzerland

[2]Integration Environment & Energy, GPO Box 8975, EPC 2736
Kathmandu, Nepal

Introduction

Middle mountain catchments in the Hindu Kush-Himalayan region are the heavily populated and mainly rainfed headwaters of many smaller, but not less important rivers, and the tributaries of the larger North-South rivers draining the mountain system. The catchments are in a fragile and vulnerable region from the perspective of water scarcity, flooding and soil erosion. It is these catchments which caused the theory of the "Himalayan Environmental Degradation" (Eckholm, 1976).

Water scarcity in particular was found to be an issue of great concern within the catchments not only in Nepal (Merz et al., 2003), but also in catchments of the middle mountains of Pakistan, India and China (Merz et al., 2005). It is mainly caused by the seasonality of water resources and their improper management (Merz, 2004; Dhakal et al., 2005). From a downstream point of view flooding and soil erosion are more important as the middle mountain catchments are rather associated with their contribution to increased flood peaks and sediment loads than with decreased flow levels in the snow and glacial melt-fed main rivers originating in the high mountains. To what extent these catchments however contribute to the large scale problems of flooding downstream in the plains of the Ganges and Brahmaputra rivers, has been subject of a long discussion (Hofer and Messerli, 1997). It is also important to note that scale has a main impact on this issue as was shown by FAO (2002).

This paper does not address the issue of these large scale problems, but rather tries to highlight the processes as identified on the basis of data from the Jhikhu Khola (JKC) and Yarsha Khola catchments (YKC) in Nepal. Both catchments were the centre of a long-term study in the People and Resource Dynamics in Mountain Watersheds of the Hindu Kush-Himalayas (PARDYP) project (White and Merz, 2005).

Study Area and Methodology

The activities of PARDYP Nepal focus on the two catchments of the rivers Jhikhu Khola and Yarsha Khola. The two catchments are both located in the middle mountains of Nepal (Figure 1). The JKC is situated about 45 km east of Kathmandu on the Arniko Highway, and covers 111.4 km^2 in total area. The YKC with a size of 53.4 km^2 is located about 190 km east of Kathmandu on the Lamosangu-Jiri Road in the Dholaka District. Further information on the watersheds is compiled in Table 1.

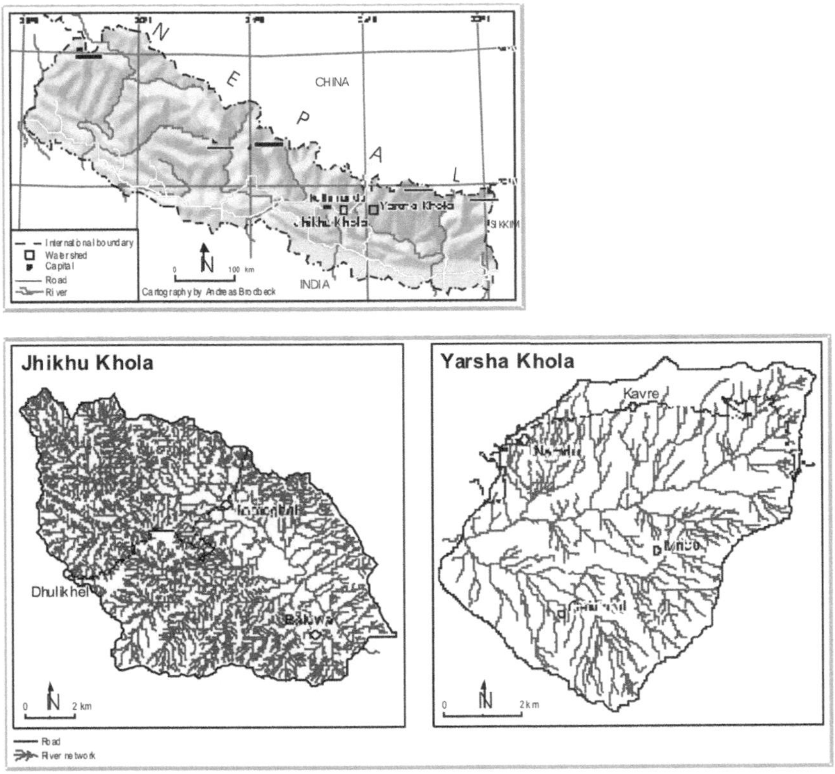

Figure 1: Location of the Jhikhu Khola (JKC) and Yarsha Khola (YKC) catchments.

Table 1: Characteristics of Jhikhu Khola (JKC) and Yarsha Khola (YKC) catchments

	JKC	*YKC*
Catchment size [km^2]	111.4	53.4
Altitudinal range [masl]	800–2200	990–3030
Physiography	Flat valley bottom of alluvial origin; short and steep slopes	Bowl shaped valley with limited flat portions
Dominant land use	Irrigated agricultural land in the valley bottom, rainfed agricultural land and forest land on the slopes	Rainfed agricultural land and forest with irrigated land along the river courses and the lower parts
General aspect	South-east (main valley extending from north-west to south-east)	South-west (main valley extending from north-east to south-west)
Annual rainfall [mm] (period) at main stations	1167–1418 (1998–2000) at Tamaghat (865 masl.)	2018–2469 (1998–2000) at Bagar (1690 masl.)
Altitudinal differentiation of rainfall [mm]	1100–1700 (1998–2000) between 830 and 1700 masl.	1600–3000 (1998–2000) between 1005 and 2640 masl.
Population density [people/km^2]	437 (1996)	386 (1996)

In both catchments a hydro-meteorological monitoring network was established according to the nested approach described in Merz (2004). Rainfall is measured with tipping buckets at event resolution, while water level is measured with floaters and pressure transducers at 30 min intervals. Discharge is measured irregularly on the basis of flow and converted to continuous time series with the help of rating curves. Sediment is sampled irregularly on the basis of flow and later converted to a sediment rating curve. Further details on the measurement and monitoring programme can be found in Merz (2004).

Results

Hydrological event analyses in relation to rainfall and erosion plot event parameters conducted by Merz (2004) basically showed that rainfall amount and rainfall intensity are the main hydro-meteorological parameters of interest. Both runoff at the plot scale as well as at the sub-catchment scale is directly correlated with these two parameters. In both catchments, the JKC and YKC, rainfall events were grouped into four clusters, which showed very good relations with the runoff behaviour at both scales. The clusters of the two catchments are compared in Table 2. In general the two systems coincide rather well and show similar minima and maxima rainfall volumes and rainfall intensities for most of the clusters. Cluster 1 in both catchments is of 2 to 10 mm rainfall volume and about 2 to 5 mm/30 min (= 4 to 10 mm/h). Rainfall volumes of cluster 2 range from about 10 to 30 mm with maximum 30 minute rainfall intensities of 3 to 15 mm/30 minutes (6 to 30 mm/h). The most important clusters for flood generation, clusters 3 and 4 differ slightly mainly due to the small number of events at this magnitude and therefore more random setting of minima and maxima. While in the JKC cluster 3 occupies the middle segment of rainfall volume and the top segment of rainfall intensity, in the YKC, the rainfall volumes of clusters 3 and 4 show the same limits. They however differ in terms of maximum intensities and show here similar lower limits as the JKC and slightly elevated maxima.

Table 2: Rainfall cluster of the Jhikhu Khola and Yarsha Khola catchments

Variable	*Cluster 1*		*Cluster 2*		*Cluster 3*		*Cluster 4*	
	Min	*Max*	*Min*	*Max*	*Min*	*Max*	*Min*	*Max*
				JKC				
P_{tot} [mm]	2.1	9.6	9.4	32.5	12.8	45.4	52.1	164.4
t_P [min]	22	250	98	728	46	421	795	1931
I_{30max} [mm/30 min]	1.8	5.4	2.7	10.4	9.4	28.7	4.7	10.7
P_{50} [%]	40.0	82.6	29.7	80.6	43.3	91.3	36.6	62.2
				YKC				
P_{tot} [mm]	3.2	10.1	12.0	30.4	20.0	69.8	26.7	69.2
t_P [min]	77	308	80.0	760.0	147.3	507.3	586.5	1121.0
I_{30max} [mm/30 min]	1.8	5.0	3.0	13.6	10.7	44.4	5.1	19.4
P_{50} [%]	20.0	79.5	6.9	92.3	20.2	94.6	10.1	68.8

Antecedent precipitation and herewith an approximation of antecedent moisture conditions only has a limited influence on the runoff behaviour at the sub-catchment scale. It further showed that the land use at the plot scale was decisive in terms of runoff generation as the land use had an impact on the soil characteristics as well as on the vegetation cover. In order to extend the spatial dimension and determine the impact of land use and other catchment characteristics, the hydrological parameters are put into relation with a number of selected characteristics. On the basis of literature and process understanding a number of catchment characteristics were identified, which may have an impact on the hydrological behaviour of the catchment during flood events. To verify these assumptions, selected characteristics were tested for their impact on the hydrological event parameters in meso-scale catchments of the middle mountains in Nepal. The hydrological event characteristics that were included in these analyses are the median runoff coefficient α [%], the median total event runoff Q_{tot} [mm], the median direct event runoff Q_E [mm] and the median peak event runoff Q_{Emax} [mm]. Q_{max}, the maximum event discharge, though probably the most important parameter in terms of flooding, was excluded in this comparative analysis as it is directly correlated to catchment area. It was replaced by Q_{Emax}, which accounts for the different catchment areas and can therefore be directly compared amongst catchments of different sizes.

These characteristics were always tested both for the median of all events as well as the median of the 10 largest events at each hydrological measurement site. As the variables are not normally distributed, linear regression cannot be used to determine a linear relation between the parameters. For this purpose the Spearman correlation coefficient was used instead. Some plots with linear regression lines are presented purely to visualize the relationships.

Morphometric and Topographic Characteristics

The areal morphometric characteristics that were tested included the catchment area, the width/elongation ratio and the drainage density (only for the JKC and its sub-catchments). The topographic characteristics that were included were the mean Topoindex, the mean relative contributing area (only JKC and sub-catchments), the mean slope and the ratio between areas below 5 degree slope and the areas of more than 15 degree slope. Elevation was not included in this analysis as the influence of elevation is largely included in the rainfall characteristics (see Merz, 2004). Table 3 presents the Spearman correlation coefficients for these catchment characteristics in relation with the selected discharge event parameters. Generally the correlations are weak and/or insignificant. However, a number of correlations can be observed. These include for the median of all events:

- drainage density and α with an r of 1 and a correlation significant at 0.0% level;
- drainage density and Q_{Emax} with an r of 0.90 and a correlation significant at 0.04% level; and
- Topoindex and Q_{tot} with an r of 0.93 and a correlation significant at 0.01% level.

In addition to these significant and strong correlations the Topoindex shows a rather strong correlation at the significance level of 0.15% with Q_E and the relative contributing area with Q_{tot}. In Figure 2 the significant and strong relationships mentioned above are visualized. It is important to note that these relationships are only for the sub-catchments of the JKC as drainage density could not be calculated for the YKC due to differently detailed mapping of the drainage network in the two catchments. The relationship between the Topoindex and the Q_{tot} additionally includes the value from the YKC.

Table 3: Spearman correlation coefficients '*r*' for morphometric and topographic catchment characteristics in relation to hydrological event characteristics

Parameters		*Median of all events*				*Median of 10 largest events*		
		α	Q_{tot}	Q_{Emax}	Q_E	Q_{tot}	Q_{Emax}	Q_E
Catchment area	*R*	0.14	–0.08	–0.52	0.08	0.42	–0.08	0.25
	Sig.	0.79	0.83	0.15	0.83	0.27	0.83	0.52
Width/elongation	*R*	–0.32	0.32	–0.62	–0.09	–0.09	**–0.74*****	–0.50
	Sig.	0.68	0.53	0.19	0.87	0.89	**0.10**	0.31
Mean slope	*R*	–0.12	–0.23	–0.03	–0.30	–0.25	–0.20	–0.25
	Sig.	0.83	0.56	0.93	0.44	0.53	0.60	0.51
Topoindex	*R*	0.74	**0.93***	0.44	0.73***	**0.75****	0.23	0.58
	Sig.	0.26	**0.01**	0.39	0.10	**0.08**	0.66	0.23
RCA	*r*	–0.60	–0.80***	–0.30	–0.60	–0.50	–0.10	–0.50
	Sig.	0.40	0.10	0.62	0.29	0.39	0.87	0.39
Slope ratio	*r*	–0.49	–0.17	–0.15	0.23	0.38	0.50	0.43
	Sig.	0.33	0.67	0.70	0.55	0.31	0.17	0.24
Drainage density	*r*	**1.00***	–0.10	**0.90***	0.00	0.40	**1.00***	0.40
	Sig.	**0.00**	0.87	**0.04**	1.00	0.51	**0.00**	0.51

* Correlation is significant at the 0.05% level (Sig.<0.05%)

** Correlation is significant at the 0.1% level (Sig.<0.1%)

*** Correlation is significant at the 0.15% level (Sig.<0.15%)

For the median of the 10 largest discharge events at all sites drainage density again shows a high and significant correlation with Q_{Emax} and the Topoindex with Q_{tot}. An additional significant and strong correlation is observed between the ratio of catchment width and catchment elongation in relation to Q_{Emax}. For the assessment of the catchment's susceptibility to floods the use of the Topoindex is therefore proposed. The width/elongation ratio can also be used. Drainage density would be an informative and good variable. Due to the difficulties in assessing this value across the region with data of different origin, it is however suggested to refrain from using it for the comparative analyses. This is also the reason why the concept of the relative contributing areas has to be dropped for these analyses.

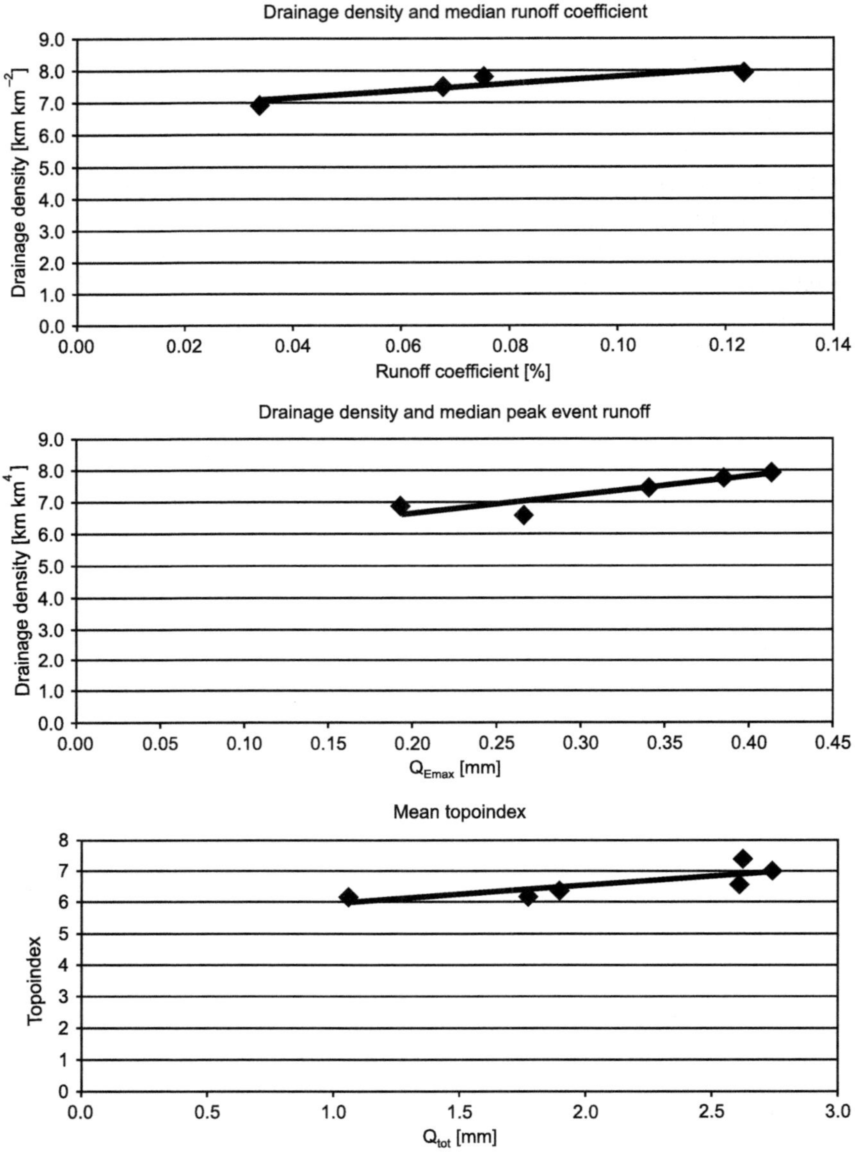

Figure 2: Linear relationships between selected morphometric and topographic catchment characteristics and hydrological event parameters.

Land Use Characteristics

For the assessment of the impact of land use on hydrological event parameters, the percentage of each land use, the ratio of cultivated to uncultivated land and the ratio of rainfed to irrigated land were used as parameters. A number of significant and strongly correlated relations were observed for these characteristics in relation to all flood events (Table 4). The strongest and most significant correlations are observed between the ratio of rainfed and irrigated agricultural land and the hydrological parameters α and Q_{tot}.

In addition, Q_{Emax} and grassland show a correlation at a significance of 0.22%. Shrub land shows a significant and strong correlation with Q_{Emax} as well as Q_{tot}. The correlation with α is significant at the 0.16% level. The other areas in the catchments, including settlements, landslides, gullies and the like, show a strong a significant correlation with α and Q_{tot} as well as a rather strong correlation with Q_{Emax}. The ratio between cultivated and uncultivated land shows very strong and significant correlation with the runoff coefficient. At the same time the correlations with Q_{tot} and Q_{Emax} are not as strong, but still quite significant.

Table 4: Spearman correlation coefficients *r* for land use related catchment characteristics in relation to hydrological event characteristics

Parameters		*Median of all events*				*Median of 10 largest events*		
		α	Q_{tot}	Q_{Emax}	Q_E	Q_{tot}	Q_{Emax}	Q_E
Cultivated/ uncultivated	*R*	**–0.77****	–0.57***	–0.50	–0.17	–0.12	0.12	0.03
	Sig.	**0.07**	0.11	0.17	0.67	0.77	0.77	0.93
Rainfed/irrigated	*R*	**–0.93***	**–0.88***	–0.46	–0.44	–0.11	0.32	0.04
	Sig.	**0.01**	**0.00**	0.21	0.23	0.78	0.40	0.92
Irrigated area	*R*	**–0.81***	–0.28	**–0.66****	–0.12	0.08	–0.07	–0.01
	Sig.	**0.05**	0.47	**0.05**	0.75	0.85	0.86	0.98
Rainfed area	*R*	–0.66	–0.47	–0.28	–0.12	–0.08	0.15	0.10
	Sig.	0.16	0.21	0.46	0.77	0.83	0.70	0.80
Forest area	*R*	0.09	0.15	0.10	–0.18	–0.15	–0.17	–0.27
	Sig.	0.87	0.70	0.80	0.64	0.70	0.67	0.49
Grass land area	*R*	**0.77****	0.48	0.23	0.52	**0.72***	0.27	0.53
	Sig.	**0.07**	0.19	0.55	0.15	**0.03**	0.49	0.14
Other areas	*R*	**0.77****	**0.65****	0.45	0.13	–0.20	–0.28	–0.32
	Sig.	**0.07**	**0.06**	0.22	0.73	0.61	0.46	0.41
Shrub area	*R*	0.66	0.58***	**0.65****	0.42	0.05	–0.13	0.07
	Sig.	0.16	0.10	**0.06**	0.27	0.90	0.73	0.87

* Correlation is significant at the 0.05% level (Sig.<0.05%)

** Correlation is significant at the 0.1% level (Sig.<0.1%)

*** Correlation is significant at the 0.15% level (Sig.<0.15%)

Four of the above correlations are visualized in Figure 3. The linear trend line is only presented for the purpose of visualization of the linear relationship and by no means to be thought of as prediction model. The two relations showing the impact of agricultural land in general and irrigated land in particular on the total event runoff and the peak event runoff, respectively, have a decreasing trend, i.e. the more cultivated land the lower the Q_{tot} and the more irrigated land the lower the Q_{Emax}. Shrub land and other land uses or covers show the opposite trend with increasing Q_{tot} and Q_{Emax} on increasing percentage of shrub and other land use/cover.

The only strong and significant correlation observed in the case of the 10 largest events is established for the percentage of grassland and the total event runoff Q_{tot}. An increase in grassland leads to an increase in Q_{tot}. Though these relationships have to be considered with caution due to the correlation problem mentioned above and the small sample number, the determined relationships seem plausible. They seem plausible on the basis of the observations on the erosion plots, where degraded land and grassland produce considerably more runoff on aggregated as well as on event basis than rainfed agricultural land. They also seem plausible on the basis of the fact that irrigated land, with its level terraces, is designed to keep water back and an enhanced water storage effect for rainfall is therefore not surprising.

The results also seem plausible on the basis of published results in the literature. For the estimation of design floods on the basis of catchment characteristics in Switzerland, Duester (1994) based his calculations on elongation factor, mean slope, areal precipitation, area of grassland and others, relative contributing areas. Elongation, here expressed as width/elongation ratio, and the area of grassland and others were also established as potential factors influencing the hydrological event characteristics. Instead of mean slope, which only shows low correlations, the Topoindex, also a product of the slope conditions in the catchments, showed high correlations with the hydrological event parameters.

In general, the correlations between the median values of the hydrological event parameters and the catchment characteristics are stronger and more significant. This observation would support the conclusions of Dangol et al. (2002). They concluded that the biggest events are a function of climatological parameters and the human influence through different land use would be negligible during these events. They also said that during minor and intermediate events human impact could be observed. These observations are presumably true for the process of flood generation in the rural context as well as natural channels. In-channel changes, e.g. bridge construction, embankments etc., may have considerable impact on the flood behaviour as for example shown by Hofer (1998).

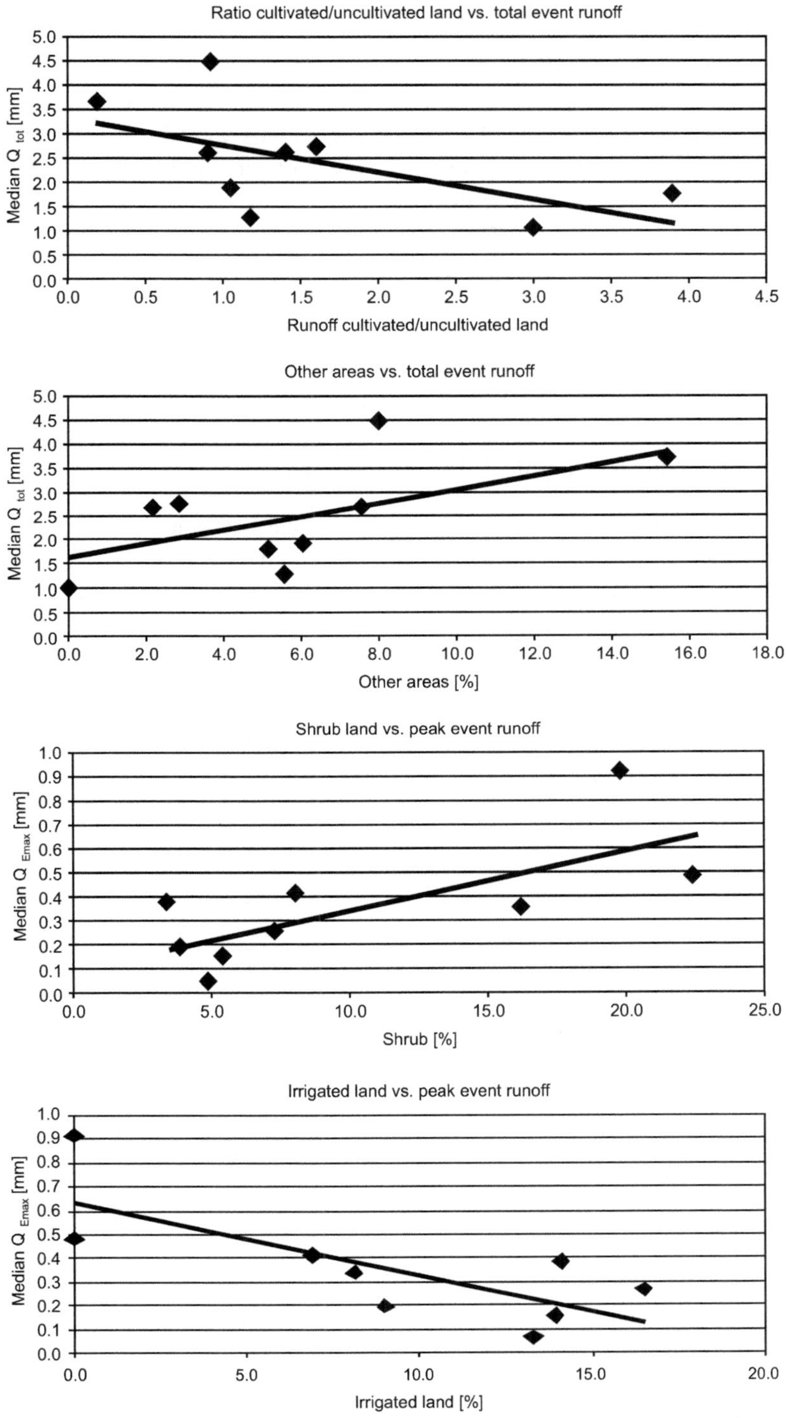

Figure 3: Linear relationships between selected land use catchment characteristics and hydrological event parameters.

Sediment Transport

On the basis of the seasonally disaggregated sediment rating curves presented in Merz (2004), seasonal sediment loads were calculated for the different sub-catchments (Table 5). In the JKC, the highest loads were estimated for sub-catchment 2 with about 34 t/ha during the pre-monsoon and monsoon seasons. This load can be explained with the large degraded area, which makes up about 12% of the total catchment area and is located in close proximity to the outlet of the catchment. This suggests that the deposition possibilities for the sediment mobilised in these degraded areas are limited and most of it is washed out of the catchment. These sediment loads from sub-catchment 2 are followed by the sediment loads from site 8, with about 24 t/ha during the pre-monsoon and monsoon seasons. Suspended sediment load is observed to be about 19 t/ha for the two seasons at the outlet of the catchment. The lowest figures are estimated for site 7, the upland sub-catchment.

Table 5: Seasonal sediment loads of the JKC and YKC (mean±standard deviation)

Site	*Catchment area [ha]*	*Pre-monsoon [t/ha/y]*	*Monsoon [t/ha/y]*	*Sum [t/ha/y]*	*Carver (1997)*
		JKC			
Site 1[1]	11141	1±1	18±2	19±2	11±1
Site 2[1]	539	2±3	32±17	34±20	15±5
Site 7[2]	74	3±4	10±5	13±9	17±11
Site 8[2]	178	9±1	15±11	24±12	-
		YKC			
Site 1[3]	5338	14±4	33±15	37±19	
Site 2[3]	1737	1±1	14±5	15±5	
Site 5[3]	32	-	18±3	-	
Site 7[3]	208	5±3	22±7	27±10	

[1] On the basis of 1993 to 1999 data
[2] On the basis of 1997 to 1999 data
[3] On the basis of 1998 to 2000 data

In the YKC, the highest loads are estimated at site 1, the outlet of the catchment followed by the loads at sites 7 and 5, both located in the same sub-catchment, and finally site 2. The reason for the highest loads at the outlet are believed to be the very steep and often from gullies dissected lower slopes of the south-facing part of the YKC as well as the quite extensive streambank erosion along many streams of this slope.

The values presented for the JKC and its sub-catchments differ from Carver (1997). However, Carver's study period was during the driest time of the study period between 1992 and 1994. By averaging the two first years of the annual sediment loads as calculated on the basis of the sediment rating curves established for this study a mean sediment load of 16 t/ha during the pre-monsoon and monsoon seasons was estimated for site 2 in contrast to a value of 15 t/ha for the two seasons by Carver (1997). At site 1 the average estimate proposed by this study for 1993 and 1994 is 12 t/ha for the pre-monsoon and monsoon seasons compared to 11 t/ha by Carver (1997). For site 7 no estimate was proposed for the years prior to 1997 as the discharge data was not adequate to produce a rating curve.

The sediment loads shown in Table 5 compare with other studies from the region as follows:

- Galay et al. (2001) compiled the sediment yields of a number of small catchment studies in Nepal. Two of the catchments are of the size of the Jhikhu Khola catchment, i.e. the Kulekhani catchment (12,500 ha), which had a sediment delivery of 20.5 t/ha/y, and the Harpan Khola (12,000 ha) showed a sediment delivery of 8.9 t/ha/y. The Bagmati at Sundarijal (1,553 ha), comparable to the Lower Gopi Khola in the Yarsha Khola catchment showed a sediment delivery of 13 t/ha/y. The Godavari catchment only showed 3 t/ha/y with an area of 1,231 ha.
- Sharma (1988) reports 45 t/ha/y for the entire Sun Kosi system with a catchment area of 19,230 km^2.

These few figures show that the data from the YKC and the JKC are plausible and within range of the other studies in the country.

The sediment loads calculated above were related to selected catchment characteristics using the correlation coefficient according to Spearman (Table 6). In general the correlations are weak and insignificant. For the pre-monsoon season sediment yields, only the Topoindex showed significant correlation at the 10% level. For the monsoon season and the annual sediment yields grassland and the ratio of cultivated to uncultivated land showed significant correlations. While grassland showed a positive correlation, which means the more grassland the higher the sediment yield, the cultivated/uncultivated ratio showed a negative correlation suggesting that an increase in cultivated land leads to lower sediment yields. This result is rather interesting if compared with the plot results of the YKC, where rainfed land produces much more sediment than grassland. This discrepancy can be explained in different ways:

- The correlation observes a splur correlation, or
- The grassland plots observed in the YKC are not representative for the soil loss generally observed on grasslands. Literature suggests 10 to 200 t/ha for degraded grassland, while well-managed pasture shows values of 0 to 10 t/ha. Joshi and Negi (2002) for example observed higher soil loss on grassland than on shrub land. In case of high stocking densities soil loss can also increase, or

Table 6: Correlation coefficients according to Spearman of sediment yield per unit area with selected catchment characteristics

		Pre-monsoon	*Monsoon*	*Annual*
Catchment area	*r*	-0.29	0.35	0.32
	Sig.	0.54	0.36	0.48
Irrigated land	*r*	-0.14	-0.18	-0.11
	Sig.	0.76	0.65	0.82
Rainfed land	*r*	0.21	-0.65*	-0.57
	Sig.	0.65	0.06	0.18
Forest land	*r*	-0.18	0.38	0.46
	Sig.	0.70	0.31	0.29
Grass land	*r*	0.11	0.77**	0.75**
	Sig.	0.82	0.02	0.05
Shrub land	*r*	-0.04	0.28	0.00
	Sig.	0.94	0.46	1.00
Other land use	*r*	0.14	0.25	0.14
	Sig.	0.76	0.52	0.76
Ratio cultivated/uncultivated	*r*	0.00	-0.77*	-0.75**
	Sig.	1.00	0.02	0.05
Ratio rainfed land/irrigated land	*r*	0.11	-0.44	-0.29
	Sig.	0.82	0.23	0.54
Degraded land	*r*	-0.60	0.70	0.80
	Sig.	0.40	0.19	0.20
Mean slope	*r*	0.36	-0.18	-0.11
	Sig.	0.43	0.64	0.82
Topoindex	*r*	-0.95*	0.56	0.21
	Sig.	0.05	0.32	0.79

r = correlation coefficient according to Spearman, Sig. = significance levels
* Correlation is significant at the 0.05% level (Sig.<0.05%)
** Correlation is significant at the 0.1% level (Sig.<0.1%)
*** Correlation is significant at the 0.15% level (Sig.<0.15%)

- The sediment mobilised on the rainfed agricultural land is successfully kept on the slopes by the lower terraces or the irrigated land (e.g. shown by Carver, 1997, who showed accumulation of soil on the irrigated terraces), or
- The high runoffs generated on the grassland are responsible for increased streambank erosion and herewith increase the total catchment sediment outputs.

Conclusive answers cannot be provided at this stage. It can be barely suggested to test the above relations with a larger sample from other middle mountain catchments and additional observations on grasslands.

Conclusions

Rainfall events in both catchments can be clustered in four clusters with low, medium, high intensity and large events. The cluster centres and their limits are tabulated above. On the catchment and sub-catchment scale the area of the catchment under grassland, degraded land and other land uses shows an enhancing effect of the floods, while the cultivated land and irrigated land in particular tend to dampen the flood peaks as well as the average event parameters. This would suggest that the terracing by the Himalayan farmers actually reduce the flood peaks, rather than, as often postulated increase them. Amongst the topographic and morphometric parameters the drainage density and the Topoindex show an increasing effect.

Sediment transport was discussed by means of sediment loads, which showed clear seasonal differences with higher sediment concentrations during the pre-monsoon season. Overall the sediment concentrations tended to decrease with catchment area, so that the highest concentrations were observed in the small upland catchments. While grassland showed a positive correlation with sediment load, which means the more grassland the higher the sediment yield, the cultivated/uncultivated ratio showed a negative correlation suggesting that an increase in cultivated land leads to lower sediment yields.

It is important to note that the relations discussed are tentative and have to be confirmed with larger samples, e.g. in comparison with the PARDYP catchments in China, India and Pakistan or other meso-scale catchments in the region. This is particularly necessary, as the correlation between two parameters does not yet explain their causal correlation (Sachs, 1997). He uses the example of the strongly positive correlation between the number of storks and the number of new borns, which obviously is a spurious correlation and does not show any functional relation.

Acknowledgements

The PARDYP project is jointly financed by the Swiss Agency for Development and Cooperation (SDC), the International Development Research Centre (IDRC) and the International Centre for Integrated Mountain Development (ICIMOD). The authors would like to acknowledge the continuous support of the donors and the continuous assistance of the colleagues in PARDYP Nepal.

References

Carver, M., 1997. Diagnosis of Headwater Sediment Dynamics in Nepal's Middle Mountains: Implications for Land Management. Unpublished PhD Thesis. University of British Columbia, Vancouver.

Dangol, P.M., Merz, J. and Weingartner, R., 2002. Flood generation in the middle mountains of Nepal. *In:* Spreafico, M. and Weingartner, R. (Eds.) Proceedings of the International Conference on Flood Estimation, March 6-8, 2002, Bern.

Dhakal, M.P., Dongol, B.S., Dongol, P.M., Merz, J. and Weingartner, R., 2004. Issues and options in water management: A case study from two middle mountain watersheds in Nepal. Paper presented at the Mountain Hydrology Conference, Berchtesgaden, Germany, September 2004.

Duester, H., 1994. Modellierung der raeumlichen Variabilitaet seltener Hochwasser in der Schweiz. Geographica Bernensia, **G44**. Geographisches Institut, Universitaet Bern, Bern.

Eckholm, E., 1976. Losing Ground: Environmental Stress and World Food Prospects. W.W. Norton & Company, New York.

FAO, 2002. Land-Water Linkages in Rural Catchments. *FAO Land and Water Bulletin,* **9.** FAO, Rome.

Galay, V., Schreier, H. and Bestbier, R., 2001. Himalayan Sediments: Issues and Guidelines. CD-ROM. Canadian International Development Agency and Canadian International Water and Energy Consultants, Vancouver.

Hofer, T., 1998. Floods in Bangladesh – A Highland-Lowland Interaction? *Geographica Bernensia,* **G48**. Institute of Geography, University of Bern, Bern.

Hofer, T. and Messerli, B., 1997. Floods in Bangladesh – Process Understanding and Development Strategies. Institute of Geography, University of Bern, Bern.

Joshi, V. and Negi, G.C.S., 2002. Rainfall characteristics, overland flow and soil loss under three vegetation types in the Himalayan mountains. *Hydrology Journal,* **25(2 & 3):** 55–64.

Merz, J., 2004. Water Balances, Floods and Sediment Transport in the Hindu Kush-Himalayan Region. Geographical Bernensia, **G72**. Department of Geography, University of Bern, Bern and International Centre for Integrated Mountain Development, Kathmandu.

Merz, J., Dangol, P.M., Dhakal, M.P., Dongol, B.S., Gao, F., Jehangir, M., Joshi, B.K., Ma, X., Nakarmi, G., Prajapati-Merz, B., Salam, A., Satyal, G.S., Shrestha, S., Verma, P.K., Weingartner, R. and Zokaib, S., 2005. Water related key issues in meso-scale catchments of the Hindu Kush-Himalayas. *In:* Stocking. M., Helleman, H. and White, R. (Eds) Renewable Natural Resources Management for Mountain Communities. International Centre for Integrated Mountain Development. Kathmandu.

Merz, J., Nakarmi, G., Shrestha, S.K., Dahal, B.M., Dangol, P.M., Dhakal, M.P., Dongol, B.S., Sharma, S., Shah, P.B. and Weingartner, R., 2003. Water: a scarce, resource in rural catchments of Nepal's Middle Mountains. *Mountain Research and Development,* **23(1):** 41-49.

Sachs, L., 1997. *Angewandte Statistik,* 8th Edition. Springer Verlag, Berlin.

Sharma, C.K., 1988. Natural Hazards and Man Made Impacts in the Nepal Himalaya. Pushpa Sharma, Kathmandu.

White, R. and Merz, J., 2005. People and Resource Dynamics in Mountain Watersheds of the Hindu Kush Himalayas Project (PARDYP) – a research for development network in mountain watersheds. *In:* Stocking. M., Helleman, H. and White, R. (Eds). Renewable Natural Resources Management for Mountain Communities. International Centre for Integrated Mountain Development, Kathmandu.

15

Inventorisation of Environmental Risk Associated with Hazardous Waste Generated in Small Scale Industrial Area of Delhi, India

Preeti Saxena and Asim K. Bhattacharyya[1]
Norwegian Institute of Air Research, P.B.100, NO-2007, Kjeller, Norway
[1]School of Environmental Sciences, Jawaharlal Nehru University
New Delhi-110067, India

Introduction

India with a total population of over a billion, has an urban population of almost 275 million, growing at about 3.5% and generating approximately 30 million tons of urban solid waste (Sashi, 2003). It may be hard to accept but true to fact the government's policy to promote small-scale sectors in India has a fatal toxic metal dumping in densely inhabited north Delhi region (Terivision, 2000). It is an emerging environmental problem in developing countries. Small scale industries have been regularly dumping extremely hazardous waste mainly fatal toxic metals against the rule of 1991 (Man and Handling, 1989) in North Delhi region every year.

Urban industrial waste is a major problem responsible for human health. These industrial pollutants are mainly discharged from metal processing industries, steel industries, chemical industries, oil refineries, sugar factories, textiles, distilleries etc. These industrial waste is producing many chemical pollutants consisting of calcium salts, many heavy metals like iron, manganese, chromium, nickel etc. and several toxic volatile elements like arsenic, selenium, mercury, lead, cadmium etc.

Along with above problems, pathogenesis of waste should also be considered because many of pathogens, germs, viruses, bacteria, fungi are depositing in this industrial solid waste and can infect people due to long persistence of untreated waste in nature. Exposure to hundreds of biological agents—viruses, bacteria, parasites, fungi and moulds—occurs in many occupational environments from industrial area to domestic places. Many bacterial pathogens especially tuberculosis and chronic parasitic diseases are some of the most common occupational diseases resulting from such exposures. According to World Health Organisation (2002) in between 10 and 30% of the workforce in industrialized countries and up to 80% in developing and newly industrialized countries are exposed to a variety of these potential hazards.

In Delhi city according to National Productivity Council, 1998, there are 1124 hazardous waste generating units in the nonconforming areas of Delhi. Approximately 70% of total hazardous waste in Delhi are from metal finishing industries (DPCC, 2001). As per NCAER (2001), 13,785 industrial units are presented in 28 (Figure 1) approved industrial areas in Delhi. Out of these 27% are polluting units which generate 151.6 tons HW/day. Approximate 70% hazardous waste of total solid waste is from metal finishing industries in Delhi (DPCB, 2001).

Wazirpur industrial area (WIA) is one of the most important one among these 28 industries. Here small and tiny scale nonferrous industries (e.g. steel, copper) factories were dumping their heavy metal rich waste and acids into less deep pools or drains. W.I.A. is a cluster of small scale industries in which more than 80% are pickling and rolling industries and other metal finishing industries (Figure 2). These are releasing a large amount of solid waste which are directly dumped on road side (Figure 3). This roadside dump is continuously mixed with area sewage and drain with the discharge of resident and slum dwellers of nearby areas. Therefore industrial sludge collected from these areas are even more dangerous than pure solid industrial waste to dispose these waste in soil. The acidic liquid waste is emitted into the drain and flowing outside. If this waste water is properly treated in effluent treatment plant then it will mix with the natural water resources, ground water, ponds through irregular drainage pattern of Delhi and subsequently into the river Yamuna, which is a major source of potable water in Delhi, thus posing a potential health and environmental risk to the people living in Delhi and downstream. The Yamuna catchment area of the National Capital Territory is one percent of the river's total catchment area but it generates more than 50% of the pollutants found in the Yamuna in its 22 km crucial stretch in Delhi. The hazardous wastes are being disposed off indiscriminately on open land or along with municipal solid waste within or outside the premises of units and contaminating the environment. Toxic waste also gets leached down upto the surface water table influencing the groundwater quality. In this article, we summarize the environmental health risks due to (i) physicochemical characterization of industrial waste collected from roadside dump with seasonal variation, (ii) distribution of heavy metals in waste water collecting in drains, (iii) heavy metals contamination in groundwater supply of this area, (iv) biological contamination associated with waste, and (v) health risk due to toxic chemicals produced from these industries.

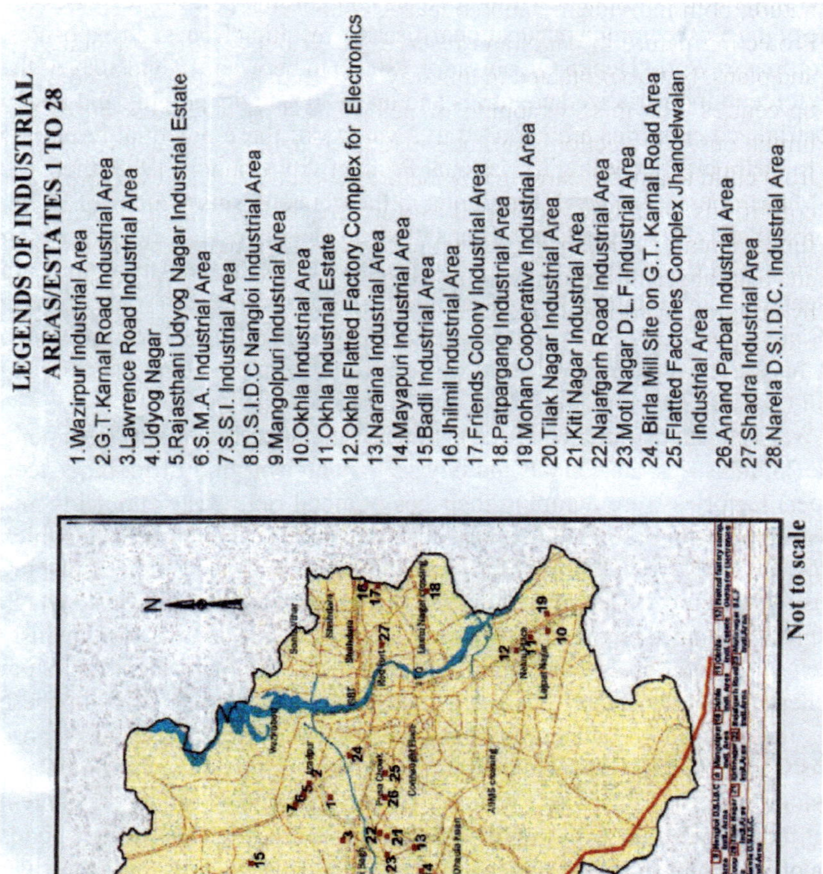

Figure 1: Approved industrial sites in Delhi.

Figure 2: Pickling and rolling industries: (a) Pickling in acid bath, (b) Hot and cold rolling, (c) Cold rolling and (d) Hot rolling.

Figure 3: Industrial waste on road side dump, drains and common dumping sites.

Study Area

Wazirpur industrial area is located on N-W part of Delhi (Figure 1). It consists of three blocks A, B and C. From the industries a large amount of solid waste is directly dumped on roadside area outside the premises and then collected in common dumping site in each block once a week. This roadside dump continuously mixes with area sewage and drain with the discharge of resident and slum dwellers of nearby areas. These sampling sites are also carrying the storm water from the industrial premises, which further connect to bigger drains in the city and finally go into river Yamuna. In addition samples were collected from river Yamuna from Wazirabad barrage and Okhla barrage; these two sites covers the 22 km polluted stretch of river Yamuna in Delhi. In addition samples were collected from Wazirabad treatment plants and 12 tube wells around this area to check the groundwater pollution.

Material and Methods

Sample Collection and Preservation

Ninety industrial solid waste samples were collected in each season, monsoon, winter and summer from July 2000 to May 2001, 30 samples in each block i.e. from ten sampling sites in each block in each specific position. 30 waste water samples were collected from drains of this area, 10 samples in each block from ten sampling sites. Waste water samples were collected in pre-treated airtight 125 ml plastic bottles and brought to the laboratory, where they were kept at 4°C until processing and analysis. Groundwater samples from eight bore wells were collected from this area. 20 samples from river Yamuna were collected from Wazirabad and Okhla barrage. The water samples were preserved with 1 ml of concentrated nitric acid till heavy metals analysis was done. The river water and groundwater samples were also treated with nitric acid and stored at 4°C. For the physicochemical examination solid waste samples were air dried, sieved (2 mm) and then preserved at room temperature with proper labelling. Water samples were filtered from 0.45 μm Millipore membrane filters and collected for the suspended material.

Procedure

Physicochemical examinations: pH, EC, CEC, organic carbon and available nitrogen in industrial wastes were estimated as mentioned in Black (1965). pH was measured immediately by 1:10 dilution by digital pHmeter. Electric conductivity of samples were measured in 1:10 dilution by digital conductivitymeter in mmho/cm. Cation exchange capacity measured by ammonium acetate method and usually expressed in milli equivalents per gram of soil. Organic carbon measured by Walkley and Black (1934) rapid titration method. In this organic matter waste gets oxidized by chromic acid and untreated dichromate is determined by back titration with ammonium ferrous sulfate (redox titration). Available nitrogen is measured by Subbiah and Asija (1956) method. This procedure involves distilling the sample with alkaline $KMnO_4$ solution and determination of the ammonia liberated which serves as an index of the available mineralizable nitrogen status.

Heavy metal analysis: The levels of heavy metals in solid waste samples were measured using a closed vessel method. In this method first remove the hydrated interlayer cations with 1M $(NH_4)_2CO_3$ and then once with 80% methanol; heat the sample at 110°C and then cool for 5 to 10 min in a desiccator; add 1 ml of aqua regia and 3 ml of HF acid in 200 mg samples; heat the samples at 110-120°C for 2-3 hours and then cool to 25°C; add 100 ml of saturated H_3BO_3 solution to the decomposed sample and immediately cap the bottle in atomic absorbtion spectrophotometer. Estimation of available trace metals was done by diethylenetriamine penta acetic acid (DTPA) as given by Lindsay & Norwell (1978). The level of heavy metals in borewell water were measured following the procedure as mentioned in standard methods for the Examination of Water and Waste Water (AWWA/APHA/WEF, 1992).

Results and Discussion

Physicochemical Characterization of Industrial Waste

The results reveal that the waste is very toxic due to low pH (Table 1) and high concentration of heavy metals (Cr, Ni, Cu). The average values of 30 samples of each block in each season are shown in Table 1. The low pH in waste is due to excessive use of acids e.g., aquaregia, H_2SO_4 and HNO_3 in the pickling and rolling industries. The sludge from pickling tank is highly acidic. Seasonal variation in pH were observed due to variation in moisture content in environment; lowest in summer due to lack of moisture in the environment and highest in monsoon due to dilution and mixing of sewage. Range of pH is 2.3-10.6 because this area is comprised of 80% pickling and rolling industries and few soap, dye, polish and textile industries; low pH is due to pickling and rolling industries and high is due to soap and other industries. The salt concentration (EC) is maximum in monsoon due to mixing of waste and almost similar in all blocks and lowest in summer.

Along with these pollutants there are many microbial nutrient found to be in high concentration and this waste can become a big source of nutrient supplier for soil. Nitrogen is found to be high i.e. 500-600 ppm in soil and organic carbon varies 3.5-4.5% which is a necessary nutrient for microbial growth. Nitrogen is high in waste because of two reasons, first high concentration of nitrate ions are coming from industrial waste water and mixing with roadside dump; second due to ammonical nitrogen coming from sewage and mixes with roadside dump. Seasonal variation in waste samples has been observed and in winter nitrogen content is highest and it may be

Table 1: Physicochemical characterisation of Wazirpur industrial waste

Parameter		*Seasons*		
		Monsoon	*Winter*	*Summer*
pH	Range	2.31-10.6	1.60-8.25	2.27-10.56
	Mean	4.53	4.88	3.969
	SD	2.052	1.94	1.677
EC (mmho)	Range	0.79-6.45	0.28-3.39	0.30-3.31
	Mean	2.30-1.33	1.68	1.195
	SD	1.405	0.742	0.723
A.N. (ppm)	Range	35-1974	61.59-1400	39.06-960
	Mean	518.50	650.37	426.56
	SD	416.67	371.81	286.405
O.C. (%)	Range	0.64-7.80	0.43-8.70	0.31-5.31
	Mean	3.60	3.347	3.636
	SD	1.874	2.504	1.632
Fe (%)	Range	7.60-22.75	8.00-24.71	9.29-25.01
	Mean	11.974	12.562	14.815
	SD	3.790	4.061	4.447
Mn (%)	Range	0.22-1.73	0.63-3.51	0.30-1.53
	Mean	0.767	1.258	0.859
	SD	0.413	0.553	0.396
Cu (ppm)	Range	201-1889	275-1445	210-1156
	Mean	488.67	475	472.87
	SD	423.45	315.76	272.32
Cr (ppm)	Range	900-12290	790-13708	780-13712
	Mean	6100.033	5680.63	5575.06
	SD	3039.037	3140.32	3137.31
Ni (ppm)	Range	60-1320	190-1109	275-1108
	Mean	626.87	565.43	589.94
	SD	242.7183	205.365	224.28

O.C.: Organic Carbon, A.N.: Available Nitrogen, CEC: Cation Exchange Capacity, EC: Electrical Conductivity, M: Monsoon, W: Winter, S: Summer.

because waste remains in original form for long period—neither diluted with drainage water nor concentrated due to evaporation. Organic carbon is high in industrial waste because it is continuously mixing with sewage and drainage water. Organic carbon in waste ranges from 0.31 to 8.70% due to high organic matter coming from sewage. Seasonal variation is observed in said waste due to dilution and proper mixing of waste. The higher value of organic carbon in summer is due to concentration of sample.

Heavy Metals in Industrial Wastes

The waste of all three blocks show high values of heavy metals in all three seasons because in pickling and rolling industries raw iron piece is processed in acids; so high percentage of iron, manganese and chromium is released. The mean values of heavy metals are found to be very high in waste as compared to Council of European Communities (Table 2). Chromium is almost five times the standards and very hazardous for plant and vegetable if this waste mix up with the soil. Nickel is almost six times the standards and also very hazardous for soil biotic factor. Copper is almost more than four times the standards of European communities. Cadmium is not found to be a problem in this waste and lead is also not found in high concentration. Iron (Figure 3) and manganese are more than six times higher than standards and cause serious risk to soil environment although iron and manganese both are very important plant nutrient if supplied in low concentration.

Table 2: Comparative analysis of heavy metals in Wazirpur industrial waste with standard recommended level (Council of European Communities, 1987)

Heavy metals	*Recommended level (ppm)*	*W.I.A. waste (ppm)*
Cu	100	478.83
Mn	1500	9500
Cr	1000	5785.29
Ni	180	594.08
Cd	20	10.60
Pb	1200	50.40

Heavy Metals in Waste Water

Mean values of pH and heavy metals in wastewater are shown in Table 3. The lowest pH in industrial wastewater in winter is 2.5. The lowest pH in this industrial area was due to the presence of a large number of rolling and pickling processing industries, where mixed acids are used to wash the rough surface of steel plates. The low pH leads to high concentrations of metals in wastewater from these industrial areas and may contribute more dissolved heavy metals to the river Yamuna (Table 3). Iron was higher in WIA (208 mgL^{-1}), which might be due to a low pH in these industrial areas and its range is 180-245 mgL^{-1}. Manganese is high in waste and its range is 35-46 mgL^{-1}. The concentration of nickel ranges 28-38 mgL^{-1} in WIA and chromium concentration was 61-71 mgL^{-1} in WIA. The concentration of copper ranges from 14.5 to 16.8 mgL^{-1}. The source of Ni and Cr in the wastewater is possibly from Ni-Cr alloy electroplating industries. There is no significant relationship between concentrations of heavy metals in different seasons i.e. winter, summer and monsoon, as heavy metals in wastewater have a continuous anthropogenic source from industries.

Table 3: Mean of pH and heavy metals (mg L^{-1}) in waste water from industrial area and water from the river Yamuna

Wazirpur Industrial	*Season*	*pH*	*Fe*	*Mn*	*Cu*	*Cr*	*Ni*
Area (*n* =10)	Mon	3.0	180	35	16.80	71.0	38
	Win	2.5	225	46	15.20	64.0	21
	Sum	2.8	248	39	14.50	61.0	27
River Yamuna at	Mon	6.75	0.56	0.30	0.16	0.12	1.02
Wazirabad barrage	Win	6.48	6.64	0.42	0.14	0.10	0.80
(*n* = 10)	Sum	6.60	0.75	0.36	0.12	0.08	0.92
River Yamuna at	Mon	7.50	0.80	0.38	0.22	0.06	0.85
Okhla barrage (*n* =10)	Win	7.20	0.86	0.61	0.20	0.05	0.58
	Sum	7.35	0.92	0.48	0.17	0.04	0.67

Heavy Metals in Suspended Material

The distribution of particulate heavy metals in suspended material is given in Table 4. High concentration of most of the heavy metals in suspended material in industrial area indicates that these industries could be emerging source of heavy metals pollution in river Yamuna. These industries are engaged in metal finishing activities and release untreated wastewater into drains, which eventually fallout in river Yamuna which is a major source of potable water in Delhi, thus posing a potential health and environmental risk to the people living in Delhi and downstream. The Yamuna catchment area of the National Capital Territory is 1% of river's total catchment area but it generates more than 50% of the pollutants found in the Yamuna in 22 km crucial stretch in Delhi. The total Yamuna before Wazirabad (Figure 4) stretch has appreciable levels of D.O., low BOD level and water is extensively used for irrigation and after Wazirabad stretch water is highly polluted due to presence of pathogens. The fecal coliforms become >5000 and Delhi region makes this count dangerously high. On the basis of water quality, the river Yamuna can be classified into four classes i.e. A, B, C and D based on D.O., BOD and fecal coliform (CPCB, 2000).

Table 4: Heavy metals (ppm) in suspended material of industrial areas and river Yamuna

	Fe (%)	*Mn (%)*	*Cu*	*Cr*	*Ni*
Wazirpur Industrial Area (*n* =30)	3.75	0.57	175	4280	330
River Yamuna at Wazirabad barrage (*n* = 10)	1.50	0.09	120	325	32
River Yamuna at Okhla barrage (*n* =10)	0.80	0.07	131	220	20

Figure 4: Water quality of river Yamuna in monsoon, post-monsoon, winter and summer. How Delhi makes the Yamuna 'dead river'. (Source: Central Pollution Control Board, April 2000—Series: ADSORBS/32/1999-2000.)

Heavy Metals in Borewell Water

Cyclic seasonal variation in temperature induces variation in bioactivity which leads to chemical changes in several parameters of river and groundwater infiltration system (L.A. Jacobes et al., 1988). Temperature variation influences the level of HMs in the final water of waste treatment plant and tubewells in Wazirpur industrial area. HMs in treatment plant samples were found in low concentration and did not exceed the regulated Indian Standards (IS) specification drinking water. HMs levels in tubewell are summarized in Figures 5-8. The level of copper, iron and manganese were found slightly more than IS standards. Nickel levels (Figure 5) ranged non-detectable level (NDL) to 0.155 mgL^{-1}. The maximum HMs concentrations were found in summer season. Copper levels exceeded the IS specification (Figure 6). The iron level (Figure 7) is higher in monsoon season in three samples but below detectable limit in summer and winter. Manganese levels were found marginally higher than IS specification in monsoon (Figure 8).

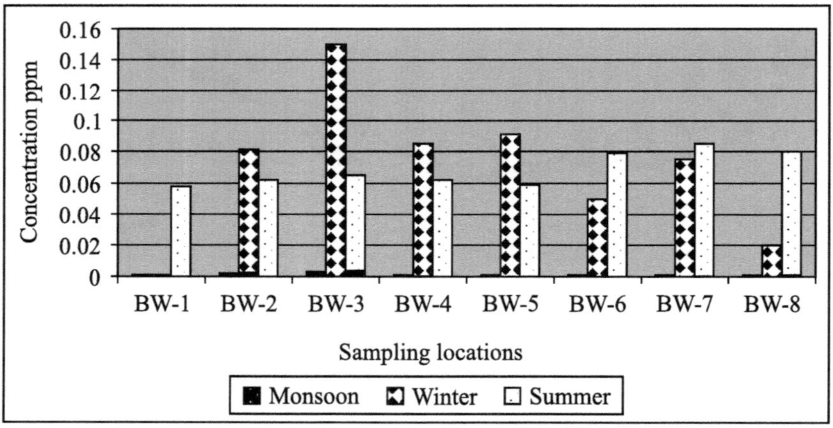

Figure 5: Seasonal variation of nickel in borewell water.

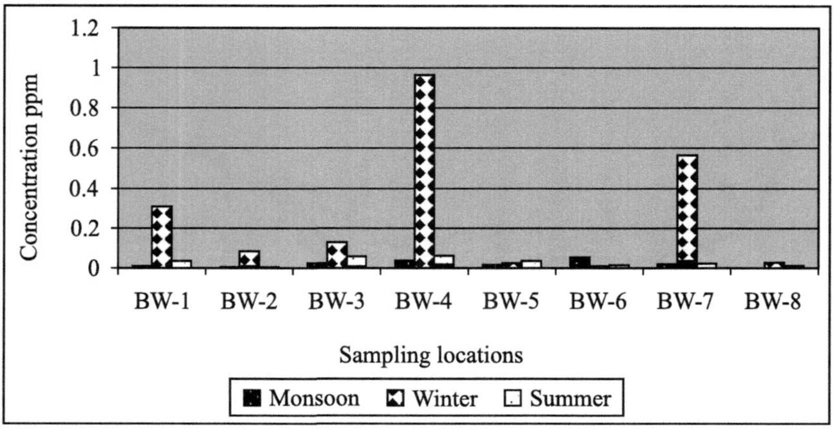

Figure 6: Seasonal variation of copper in borewell water.

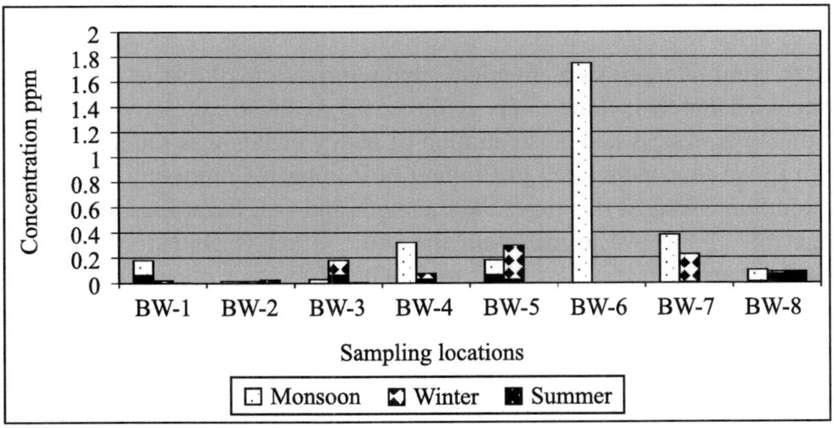

Figure 7: Seasonal variation of iron in borewell water.

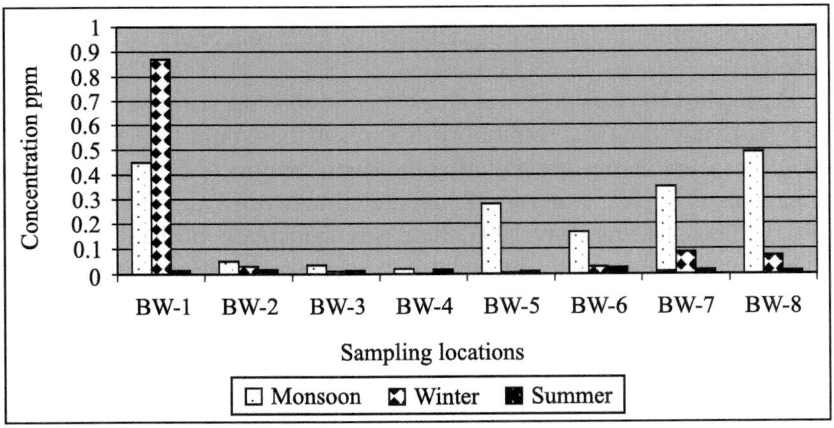

Figure 8: Seasonal variation of manganese in borewell water.

Pathogenesis in Waste

The long persistence of untreated waste in environment causes pathogenesis. These pathogens are coming from area sewage and fecal discharge of residents, slum dwellers and animals. Total 12 types of parasites were obtained in waste. *E. histolytica* is causative agent for Amoebic dysentery, *Ascaris* for ascariasis, *Trichuria* for diarrhoea, anemea and eosinophilia, *Negleria* for meningitis, Strongyloides larva for nausea, dizziness and bloody diarrhoea and hookworm for diarrhoea and loss of health and collapse. One group of intestinal bacteria are coliform bacteria present in high frequency, which is an indication that an environment has been contaminated by human sewage. In addition more than 11 types of pathogenic bacteria are present in waste which come from faeces of infected individuals. These are causative agent of typhoid, bacillus dysentery, gastroenteritis and cholera.

Conclusion

After complete physico-chemical characterization of waste it has been concluded that this waste is very acidic (2.5-3.5) due to excessive use of acids in pickling process and concentration of heavy metals was found to be very high in waste as compared to Council of European Communities (Table 2). Chromium is almost five times the standards and very hazardous for plant and vegetable if this waste mix with the soil. Nickel is almost six times the standards and also very hazardous for soil biotic factor. Copper is almost more than four times the standards of European communities. Cadmium is not found to be a problem in this waste and lead is also not found in high concentration. Iron and manganese are more than ten times higher than standards and cause serious risk to soil environment although iron and manganese both are very important plant nutrient if supplied in low concentration. This waste can be very precious if these metals can be recycled. Along with these pollutants there are many microbial nutrient found to be in high concentration and this waste can become a big source of nutrient supplier for soil.

During last 10 years there was an increase in absolute terms in the incidence rate of both intestinal and respiratory diseases in these industrial estates (GOI, 2001). These industries are polluting the river Yamuna and also transmitting the pathogens in river water in 22 km stretch between Wazirabad and Okhla (CPCB, 2000). During 1989-1997, there was an increase, in absolute terms, in the incidence rate of both intestinal and respiratory diseases in Delhi city (Pruss, 1999). However, respiratory diseases have comparatively gained more importance. The city has witnessed increasing cases of asthama, bronchitis and severe cough type symptoms. While many traditional types of intestinal diseases have been controlled, new strains causing these have emerged. The problem is primarily associated with poor sanitation and hygiene and is exacerbated by factors such as malnutrition. The dumping of industrial waste and urban waste in open lands is also a major cause of these diseases.

The results obtained recommends the regular treatment of waste water and prevent the dumping of untreated waste on open land to avoid the leaching of HMs upto the surface watertable and influencing the groundwater quality. Waste water from industrial units may be point source of heavy metals to river Yamuna in Delhi. There is a great need to install an integrated waste water treatment plant in this industrial area to avoid the pollutants entering into water bodies and save the ecosystem. A detail environmental awareness education is required for implementation of all necessary precautions during collection, transportation and disposal of these. Long persistence of these waste in field should be avoided to reduce the growth of pathogens. From the results mentioned above we can assess the health risk associated with waste which will not only be a problem of area workers but also for complete surrounding environment.

References

Bicudo, J.R. and Goyal, S.M., 2003. Pathogens and manure management systems: A review. *Environmental Technology*, **24:** 115-130.

Black, C.A., Evans, D.D., Ensminger, L.E., White, J.L. and Clark, F.E., 1965. Methods of Soil Analysis, Chemical and Microbiological Properties. American Society of Agronomy, Inc. Publication, Madison, Wisconsin.

CPCB (Central Pollution Control Board), 2000. Water Quality Status of Yamuna River, April 2000. CPCB, New Delhi, 64 pp.

Dixit, R.C., Verma, S.R., Nitnaware, V. and Thacker, N.P., 2003. Heavy metals contamination in surface and groundwater supply of an urban city. *Indian J. Environ Hlth.*, **45(2):** 107-112.

GOI (Government of India), 2001. Delhi 21: Urban Improvement and Infrastructure Improvement Project. Government of India, New Delhi.

Guan, Tat Yee and Holley, Richard A., 2003. Pathogen survival in swine manure environments and transmission of human enteric illness: A review. *Journal of Environmental Quality*, **32:** 377-382.

Lindsay, W.L. and Norwell, W.A., 1978. Development of a DTPA. Soil test for zinc, iron, manganese and copper. *Soil Sci. Soc. Am. J.*, **42:** 421-428.

MoEF (Ministry of Environment and Forests) and GNCTD (Government of National Capital Territory, Delhi), 2001. Status Report for Delhi 21. MoEF and GNCTD, Government of India, New Delhi.

NPC (National Productivity Council), 1997. Technical Environmental Impact Assessment Study for Identification of Hazardous Waste Sites in Delhi Region. NPC, Delhi.

Onibokun, A.G., 1999. Managing the Monster: Urban Waste and Governance in Africa.

Pruss, A., Giroult, E. and Rushbrook, P., 1999. Safe Management of Wastes from Health Care Activities. World Health Organisation, Geneva.

Singh, Sarman, 2000. Biological Contamination of Drinking Water in New Delhi: Isolation of free living Amoebas as a highly sensitive index of water contamination. *Water International*, **25(3):** 403-409.

Sivramkrishna, Sashi, 2003. The urban solid waste management problem in India: An economic approach and framework for policy. *Journal of Solid Waste Technology and Management*, **29(1):** 8-23.

Terivision, 2000. Fundamental Problem of Waste Generation and Disposal in Delhi, New Delhi.

Williams, S.T., Sharpe, M.E. and Holt, J.G. (Eds), 1989. Bergey's Manual of Systematic Bacteriology, Vol. 4. Lippincott, Williams & Wilkins, Baltimore.

16

The Impact of Land Use on Nutrient Concentration in Upper Streams of Waters in Slovenia

Marina Pintar, Boris Kompare[1], Urška Bremec[2], Elizabeta Gabrijelčič[2], Gregor Sluga[2], Matej Uršič[1] and Lidija Globevnik[2]

University of Ljubljana Biotechnical Faculty Agronomy Department, Jamnikarjeva 101

[1]University of Ljubljana Faculty of Civil and Geodetic Engineering, Jamova 2

[2]Institute for Water of the Republic of Slovenia, Hajdrihova 28
SI-1000 Ljubljana, Slovenia

Introduction

According to the Water Framework Directive (2000/60/EU), the EU countries have to define water bodies characterised with the same ecological, morphological and water quality characteristics. The main aim of the above mentioned directive is to reach good water status regarding quality and quantity of water by year 2015. The article 5 of the directive among others requires that every EU country makes the analysis of characteristics of watersheds and of the impacts of men's activity in them. Water bodies at risk should be defined as a first step. These are defined as water bodies for which we are not completely sure that water quality will meet the requirements by year 2015. If someone wants to improve water quality he should know how different men's activities in the watershed impact water quality. Based on this proved impact of men's activity in a catchment and water quality, proper measures in watershed for increased water quality should be determined and implemented. The most problematic situation could be if water quality is low and there is no connection between water quality and men's activity in a catchment.

Different point and non-point pollution sources can deteriorate the water quality. Agricultural activities have been defined as major sources of non-point pollutants in water (Charbonneau and Kondolf, 1993), even in some cases population with disorganised sewage systems could be the main source of non-point pollutants in water as well (Aravena et al., 1993). Agricultural land use is supposed to produce the highest amounts of nitrogen (Tong and Chen, 2002), still Fisher et al. (2000) reported that municipal sources of nutrients must be reduced too, to make significant progress in the watershed. There is a strong relationship between land use and the quality of water especially for nitrogen and phosphorus in upland watershed (Gburek and Folmar, 1999). Changing land use and land management practices are one of the main factors impacting runoff water quality (Mander et al., 1998). Current methods on predicting water quality in river catchments based on land-use patterns are still in development (Tong and Chen, 2002).

The main purpose of this work is to examine statistical relationship between land use and nitrogen and phosphorus concentration in upper part of Slovene water streams.

Materials and Methods

Land Use

We defined 126 fourth-range watersheds in Slovenia using a computer-based GIS software system Arc Info. We have chosen 35 water quality-monitoring stations for further analyses based on criteria that they are located in watersheds with the area less than 500 km^2 (Figure 1). In some 4th range watersheds there

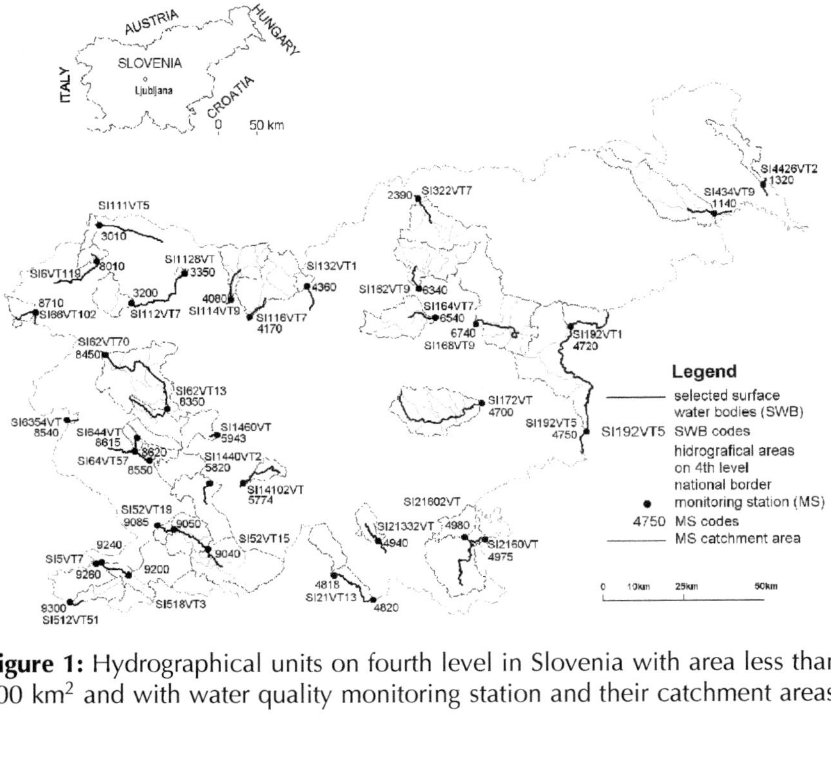

Figure 1: Hydrographical units on fourth level in Slovenia with area less than 500 km^2 and with water quality monitoring station and their catchment areas.

is more than only one water quality monitoring station. For example on water course Sotla, there are two water quality monitoring stations with station codes 4720 and 4750 (Table 1). Accurate catchment area was defined for each water quality-monitoring station.

Land use from Corine Land Cover 2000 (CLC 2000) was determined for each water quality-monitoring station catchment area. Resolution used for land use determination was 25 × 25 m^2. Land use categories used for this analysis were as follows: urban, forest, fields, orchards and vineyards, grassland and mixed land use. Once the image was classified, the area extent of each land use category was calculated for each catchment area.

Table 1: Watershed area (km^2) of water quality monitoring station and shares (%) of land use in them

Water course name	*Water quality monitoring station*			*Land use (%)*					
	Name	*ID*	*Area (km²)*	*u*	*fo*	*fi*	*ov*	*g*	*m*
Ščavnica	Pristava	1140	79	2	40	32	1	5	21
Kobiljski Potok	Mostje	1320	68	0	66	0	1	2	32
Mislinja	Otiški vrh	2390	275	2	28	27	3	2	37
Sava Dolinka	Podkoren	3010	29	2	89	0	0	8	0
Sava Bohinjka	Sv. Janez	3200	94	0	95	0	0	1	1
Jezernica	Mlino	3350	7	10	42	0	0	0	22
Tržiška Bistrica	Podbrezje	4080	142	2	87	0	0	1	10
Kokra	Kranj	4170	221	2	82	0	0	4	12
Kamniška Bistrica	source	4360	50	2	70	0	0	6	21
Mirna	Boštanj	4700	73	4	48	1	0	3	45
Sotla	Rogaška Slatina	4720	437	1	37	3	0	8	50
Sotla	Rakovec	4750	56	5	71	6	0	11	7
Kolpa	Osilnica	4818	231	2	67	1	0	13	16
Kolpa	Petrina	4820	82	0	88	0	0	8	4
Rinža	Koèevje	4940	81	0	70	0	2	6	22
Lahinja	Primostek	4975	119	0	89	0	0	6	5
Krupa	source	4980	340	1	72	4	2	3	18
Cerkniščica	Cerknica (Dolenja vas)	5774	173	5	47	1	0	16	31
Pivka	Postojna	5820	83	2	71	0	0	0	26
Logaščica	Jaèka	5943	206	5	59	1	0	18	17
Paka	Reèica	6340	172	1	64	2	2	6	24
Bolska	Dolenja vas	6540	413	3	45	1	0	9	41
Voglajna	Celje	6740	293	1	60	3	0	7	30
Soča	Trenta	8010	94	0	95	0	0	1	1

(*Contd.*)

Table 1 (*Contd.*)

Idrijca	Podroteja	8350	490	1	79	0	0	4	16
Idrijca	Hotešk	8450	87	3	75	2	0	4	16
Koren	Nova Gorica	8540	6	0	80	0	0	4	15
Vipava	source	8550	29	2	89	0	0	8	0
Hubelj	source	8615	48	0	91	0	0	3	6
Hubelj	Ajdovšèina	8620	98	0	92	0	0	0	8
Nadiža	Potoki	8710	7	10	42	0	0	0	22
Reka	Topolc	9040	423	1	77	0	0	2	20
Reka	Cerkvenikov mlin	9050	450	1	77	0	0	2	20
Reka	Matavun	9085	217	1	76	0	0	5	18
Rižana	source	9200	288	1	79	0	0	1	18
Rižana	Dekani	9240	218	1	76	0	0	5	18
Rižana	Bertoki	9260	118	1	84	0	0	3	11
Dragonja	Podkaštel	9300	191	0	78	0	0	5	16
Average			183	2	70	2	0	5	19
Minimum			7	0	28	0	0	0	0
Maximum			490	10	95	32	3	18	50

Note: u = urban area, fo = forest, fi = fields, ov = orchards and vineyards, g = grassland, m = mixed land use

Water Quality

Concentration of nitrate, ammonium and phosphate in surface water for the period 1998-2002 has been obtained from the state's monitoring programme database. In the state's monitoring, 132 sampling stations for surface water are included. The monitoring program has been started in 1986 (ARSO, 2002). We took data from the period around the year 2000, when land use data were determined. Water samples were taken once to ten times per year.

Statistical Analyses

The question about relationship between land use and water quality has been addressed by applying multiple regression method considering nutrient concentration as dependent variables and the area of land uses as independent variables. Statistical analyses were performed with Statgraphics plus software version 4 (Manugistics®). In most cases non-point pollutants received from contributing catchment area can be described in the form of an exponential model:

$$npp = \alpha e^{\left(\sum_1^i \beta_i luc_i\right)}$$

where *npp* is non-point pollutant concentration, *luc* is land use category, α is the intercept, β is a parameter that specifies the direction and strength of the relationships between each land use category and non-point pollutant concentration, i is a number of different land use category in a catchment (Basnayat et al., 1999). Following this formula, the dependent variable in the multiple regression analysis was transformed with natural logarithm to reduce asymmetric distribution. Other authors (Tong and Chen, 2002; Fisher et al., 2000) reported the need of log transformation of pollutant concentration data to normalize them and prepare for further statistical estimation.

Results and Discussion

Land Use

Catchment areas for selected water quality monitoring stations have 183 km^2 in average (range is 7-490 km^2). There is forest in all watersheds and it is the most abundant land use in watersheds in average. It covers 70% of area in average with minimum and maximum, 28% and 95% respectively. Shares of mixed land use, grassland, fields, urban land use and orchards and vineyards follow them. Observed headwater catchment areas have higher share of forest and lower share of agricultural land as they are in Slovenia as a whole, which they are, according to CLC 2000, 61% and 25% respectively. Some land use categories (for example orchards and vineyards or fields) do not exist in all watersheds. Average parcel in Slovenia covers an area less than 0.5 ha. A lot of parcels have dimensions smaller than the chosen resolution for land use determination so the share of mixed land use is quite high (Table 1).

Water Quality

Overall average concentration of ammonium, nitrate and phosphorus in surface water in observed monitoring station in the period 1998-2002 were 1.059 (<0.002-19.508), 4.087 (0.433-12.089) and 0.571 (<0.002-11.383) mg/L respectively. Pecárova and Pecár (1996) reported average annual nitrate concentration 1.76-2.78 mg/L in some mountainous catchments of Slovakian surface waters. Detailed data about pollutants concentration on monitoring stations are presented in Table 2.

Land Use and Pollutants Concentration in Surface Water Linkage

The output of simple regression analysis shows the results of fitting a linear model to describe the relationship between ammonium and phosphorus (Table 3). The correlation coefficient equals 0.82, indicating a moderately strong relationship between the concentrations of ammonium and phosphorus. Since the P-value in the ANOVA table is less than 0.01, there is a statistically significant relationship between these two pollutants at the 99% confidence level. Both of pollutants are bonded on suspended solids in water and this is the reason for strong relationship between these two pollutants. On the contrary,

Table 2: Sampling stations, number of samples and average, minimum and maximum concentrations (mg/L) of ammonium, nitrate and phosphorus in water in selected monitoring sampling stations in Slovenia for the period 1998-2000

Sampling station ID	*Number of samples*	*Concentrations (mg/L)*								
		Ammonium			*Nitrate*			*Phosphorus*		
		Average	*Min.*	*Max.*	*Average*	*Min.*	*Max.*	*Average*	*Min.*	*Max.*
1140	20	1.244	0.330	5.370	4.558	0.010	16.640	0.835	0.153	3.970
1320	5	0.060	<0.002	0.130	5.912	5.000	6.370	0.120	0.092	0.203
2390	15	0.215	0.050	0.410	6.975	5.830	8.400	0.190	0.072	0.369
3010	14	0.014	<0.002	0.150	1.847	0.630	2.400	0.009	<0.002	0.064
3200	14	0.005	<0.002	0.040	1.793	1.400	2.100	0.011	<0.002	0.087
3350	19	0.030	<0.002	0.100	0.769	0.010	1.900	0.020	<0.002	0.049
4080	20	0.090	<0.002	0.240	3.042	1.770	5.200	0.093	<0.002	0.200
4170	20	0.078	<0.002	0.360	4.284	0.059	5.600	0.133	0.059	0.319
4360	15	0.004	<0.002	0.030	2.007	1.560	2.600	0.003	<0.002	0.003
4700	18	0.041	<0.002	0.290	4.199	3.030	7.310	0.104	0.043	0.366
4720	20	1.652	0.580	8.150	4.876	0.140	8.220	1.449	0.238	4.700
4750	20	0.244	<0.002	1.920	5.096	1.890	9.500	0.346	0.136	0.954
4820	20	0.004	<0.002	0.040	3.067	2.020	4.200	0.004	<0.002	0.020
4940	9	2.593	0.250	10.320	5.202	0.100	8.490	1.396	0.218	4.560
4975	16	0.018	<0.002	0.090	4.659	2.370	5.900	0.051	0.003	0.095
4980	20	0.003	<0.002	0.030	5.146	4.200	7.200	0.054	0.032	0.080
4818	20	0.005	<0.002	0.005	3.303	2.260	3.303	0.009	<0.002	0.009

(*Contd.*)

Table 2 (*Contd.*)

5774	30	2.267	0.080	30.310	2.533	0.900	5.730	1.113	0.089	13.760
5820	16	0.433	<0.002	3.210	0.43	0.00	3.21	0.433	<0.002	3.210
5943	20	6.337	0.750	15.750	4.560	1.080	17.200	1.195	0.162	3.950
6340	20	2.704	0.600	5.490	5.950	2.350	7.980	1.076	0.257	2.560
6540	15	0.137	<0.002	0.330	12.089	9.300	15.900	0.111	0.051	0.231
6740	20	1.452	0.200	3.610	5.831	0.320	7.600	0.683	0.159	2.052
8010	7	<0.002	<0.002	<0.002	1.699	1.320	2.280	<0.002	<0.002	0.003
8350	21	0.011	<0.002	0.100	5.841	4.180	7.710	0.008	<0.002	0.043
8450	15	0.007	<0.002	0.070	3.721	2.440	4.800	0.014	<0.002	0.042
8540	6	19.508	2.210	37.200	1.050	0.200	4.000	11.383	6.400	17.660
8550	25	0.005	<0.002	0.070	5.758	4.000	8.260	0.020	<0.002	0.038
8615	25	0.007	<0.002	0.070	5.293	4.050	6.940	0.007	<0.002	0.032
8620	19	0.105	<0.002	0.730	4.542	0.030	8.070	0.304	0.087	1.490
8710	10	<0.002	<0.002	<0.002	3.864	3.280	4.600	0.010	<0.002	0.081
9040	20	0.254	0.040	0.880	3.486	2.280	5.000	0.146	0.038	0.340
9050	29	0.024	<0.002	0.130	4.455	1.780	6.100	0.069	0.003	0.147
9085	25	0.012	<0.002	0.070	3.890	0.080	6.170	0.046	0.003	0.098
9200	40	0.003	<0.002	0.040	3.584	2.600	4.890	0.038	0.003	0.249
9240	22	0.050	<0.002	0.500	3.948	2.100	5.260	0.049	<0.002	0.131
9260	16	0.604	<0.002	4.850	4.209	0.190	8.600	0.170	<0.002	1.070
9300	16	0.010	<0.002	0.040	1.833	0.200	4.500	0.006	<0.002	0.036
Average		1.059			4.087			0.571		
Minimum		<0.002			0.433			<0.002		
Maximum		19.508			12.089			11.383		

Note: <0.002 = under detection limit

nitrate is dissolved in water and there is no statistically significant relationship between nitrate and ammonium or nitrate and phosphorus although all of them came mainly from agriculture and/or from urban land use.

There is a statistically significant relationship between the shares of different land use in the watershed and ammonium, phosphorus and nitrate concentration in surface water on sampling station at the 99% confidence level. *P*-values for all three models from the ANOVA are less than 0.01 (Table 3). Stepwise regression procedure was used in the selection of independent variables for the models.

Table 3: Linear regression equation for ammonium and nitrate and multiple regression equations for pollutants concentration depending on land use in the watersheds

Ammonium (mg/L) vs. phosphorus (mg/L): r^2=67%; $P = 0.0000$ ammonium = $-0.10 + 2.36 \times$ phosphorus
Ammonium (mg/L) vs. land use (%): r^2=56%; $P = 0.0001$ ln (ammonium) = $-33.46 + 1.10 \times u + 0.29 \times fo + 0.31 \times fi + 0.27 \times g + 0.34 \times m$ (9.10) (0.28) (0.09) (0.10) (0.08) (0.09)
Phosphorus (mg/L) vs. land use (%): r^2=67%; $P = 0.0000$ ln (phosphorus) = $-30.34 + 0.91 \times u + 0.26 \times fo + 0.29 \times fi + 0.25 \times g + 0.32 \times m$ (6.40) (0.20) (0.06) (0.07) (0.06) (0.07)
Nitrate (mg/L) vs. land use (%): r^2=42%; $P = 0.0019$ ln (nitrate)= $-3.06 + 0.04 \times fo + 0.06 \times fi + 0.07 \times g + 0.05 \times m$ (1.10) (0.01) (0.01) (0.02) (0.01)

Note: u = urban area, fo = forest, fi = fields, g = grassland, m = mixed land use. Standard error of particular independent variable is in parenthesis.

The relative shares of urban areas, forest, fields, grasslands and mixed land use have an impact on the ammonium and phosphorus concentration in the surface water on the sampling station. Orchards and vineyards, which are not very abundant in the upper watershed areas, do not have a significant impact on concentration of any of the observed pollutants.

The observed upper streams, supposed to be less populated than the lowland ones in Slovenia analyses, show the most pronounced impact of urban area to ammonium and phosphorus concentration in observed surface water. Also, Fisher et al. (2000) reported that urban areas have huge impact on water quality and that municipal sources of nutrients must be reduced to make significant progress in the watershed. The relative share of urban areas has a statistically significant impact on ammonium and phosphorus concentration in the surface water, but it does not impact the nitrate concentration. The main reason for positively correlated impact of urban area to ammonium (and not to nitrate) in surface water is probably in the form of nitrogen in human excreta. Pecárova and Pecár (1996) reported relatively high phosphate concentration in Slovakian watershed resulting from contribution of municipal sewage, but wastewater was only treated mechanically. Tong (1990) also stated that urban development in the watershed could cause substantial modification on water quality.

There is no common opinion what kind of data and analyses provide good results about land use impact to surface water quality. Some authors proposed to evaluate a role of riparian buffers of undisturbed vegetation (Barling and Moore, 1994) or a role of contributing zone to a water stream (Basnayat et al., 1999). Basnayat et al. (1999) reported that land use in the entire catchment did not have any statistically significant impact on water quality; meanwhile this changed when they included in the analyses land use data of contributing area. Meador and Goldstein (2003) also proposed that only agricultural land use within a basin is too general descriptor which may not adequately characterize local activities affecting water physicochemistry. On the other hand some studies have shown that stream ecosystems can be more influenced by land use at regional or broad geographic scales and that land use may be more important than stream buffers for stream restoration (Richards et al., 1996). Perry and Vandrekilen (1996) found some correlations existing between pollution loading and land use, so there is a potential to improve water quality with proper land use management practices. However, since all our three models show statistically significant impact of land use we assumed that such a simple method is good enough to incorporate it into a first row estimation if water body is a risk or not. For more detailed, more time and money consuming investigations, some additional land use data should be taken into account, e.g. data about spatial land use distribution in catchments or distribution of buffer zones or land use data in contributing zones.

Besides these data, the data on extent and quality of sewerage systems and wastewater treatment plants shall also be taken into account. It is estimated that compliance to the EU urban waste water treatment directive employing only second-stage treatment will *increase* total nitrogen and phosphorous loads in our rivers. For nutrients reduction in the rivers, the third-stage WWTP shall be designed and operated.

Conclusions

Our research showed statistical relationship between land use categories and nutrient concentration in surface water in upper streams. To improve water quality appropriate measures regarding urban land use and agricultural land use should be implemented. Presented methodology can provide a simple but effective tool for planners and policy makers as well as government resource managers for defining the impacts of land use on water resources quality. This methodology could be incorporated into the water bodies at risk defining methodology. It could serve as a first rough estimation if particular water body is at risk or not.

References

Aravena, R., Evans, M.L. and Cherry, J.A., 1993. Stable Isotopes of Oxygen and Nitrogen in Source Identification of Nitrate from Septic Systems. *Ground Water,* **31(2):** 180-186.

ARSO (Environmental Agency of the Republic of Slovenia), 2002. Program Monitoringa Kakovosti Površinskih Voda Za Leto 2002 (Programme for Monitoring the Surface Water Quality for the Year 2002).

Barling, R.D. and Moore, I.D., 1994. Role of buffer strips in management of waterway pollution: A review. *Environmental Management,* **18:** 543-558.

Basnayat, P., Teeter, L.D., Flynn, K.M. and Lockaby, B.G., 1999. Relationships between landscape characteristics and nonpoint sources pollution inputs to coastal estuaries. *Environmental Management,* **23(4):** 539-549.

Charbonneau, R. and Kondolf, G.M., 1993. Land Use Change in California, USA: Nonpoint Source Water Quality Impacts. *Environmental Management,* **17(4):** 453-460.

Fisher, D.S., Steiner, J.L., Endale, D.M., Stuedemann, J.A., Schomberg, H.H., Franzluebbers, A.J. and Wilkinson, S.R., 2000. The relationship of land use practices to surface water quality in the Upper Oconee Watershed of Georgia. *Forest Ecology and Management,* **128:** 39-48.

Gburek, W.J. and Folmar, G.J., 1999. Flow and chemical contributions to stream flow in an upland watershed: A baseflow survey. *Journal of Hydrology,* **217:** 1-18.

Mander, U., Kull, A., Tamm, V., Kuusements, V. and Karjus, R., 1998. Impact of climatic fluctuations and land use change on runoff and nutrient losses in rural landscape. *Landscape and Urban Planning,* **41:** 229-238.

Meador, M.R. and Goldstein, R.M., 2003. Assessing water quality at large geographic scales: Relations among land use, water physicochemistry, riparian condition, and fish community structure. *Environmental Management,* **31(4):** 504-517.

Pecárova, P. and Pecár, J., 1996. The impact of land use on stream water quality in Slovakia. *Journal of Hydrology,* **180:** 333-350.

Perry, J. and Vanderklein, E., 1996. *Water Quality Management of a Natural Resource.* Blackwell Science, Cambridge, Massachusetts.

Richards, C., Johnson, L.B. and Host, G.E., 1996. Landscape-scale influences on stream habitats and biota. *Canadian Journal of Fisheries and Aquatic Sciences,* **53:** 295-310.

Tong, S.T.J., 1990. The hydrologic effects of urban land use: A case study of the Little Miami Basin. *Landscape and Urban Planning,* **19:** 99-105.

Tong, S.T.Y. and Chen, W., 2002. Modeling the relationship between land use and surface water quality. *Journal of Environmental Management,* **66:** 377-393.

17

Recovery of Headwater Catchments and Lakes Affected by the Acid Atmospheric Deposition

J. KrečEk and Z. Horicka[1]
Department of Hydraulics and Hydrology, Czech Technical University
[1]Department of Ecology, Charles University
Prague, Czech Republic

Introduction

The region of the Jizera Mountains (350 km^2, altitude of 50° 40'-50° 52', longitude of 15° 08'-15° 24', humid temperate zone, Northern Bohemia, Czech Republic) is an important water resource area with a variety of downstream benefits. In the 1980s, headwaters of the Jizera Mts have been declined as a consequence of the acid atmospheric deposition, the die-back of spruce plantations, and commercial forest practices (an extensive clear-cut and use of heavy mechanisation). Strategies of nature protection and conservation (landscape protected areas, nature reserves or protected headwater regions) collapsed because of a very limited focus. After the clear-cut of spruce plantations, *Junco effusi-Calamagrostietum villosae* became a new dominant community in headwater catchments of the Jizera Mts Reforestation of large cleared areas was complicated and the upper plain of the mountains has been overgrown by invasive grasses (particularly *Calamagrostis villosa*).

In the early 1990s, restoration of mountain watersheds in the Jizera Mts started with the political and economic changes in East Europe. The atmospheric deposition of sulphate decreased to cca 40% of the year 1987, and the issue of nature protection and landscape restoration became open to a wider discussion (Brocksen and Wisniewski, 1988; Krecek, 1994; Haigh and Krecek, 2000).

Methods

Since 1982, a long-term hydrological investigation has been conducted in the experimental basin Jizerka (area of 100 ha, elevation: 860-980 m) to study

Table 1: Characteristics of the studied reservoirs

Parameter	*Reservoir*		
	Bedrichov	*Sous*	*Josefuv Dul*
Year	1905	1915	1982
Altitude (m)	775	769	733
Maximum depth (m)	13.5	19.3	38.2
Volume (10^6 m^3)	2.1	7.6	23.3
Surface area (ha)	37	86	145
Watershed (km^2)	4.3	14.0	19.8
Residence time (days)	41	179	464

effects of clear-cut of mature spruce stands, and watersheds of three drinking reservoirs: Bedrichov, Sous and Josefuv Dul (Table 1).

To evaluate the atmospheric deposition in the open field and under the forest canopy, four forest stands (30 × 30 m, 900 m^2) were instrumented with 10 rain-gauges to collect throughfall and stemflow.

The monitoring of water quality and biota in reservoirs included regular sampling, and in situ measurement of transparency (Secchi disc), and conductivity, pH and content of dissolved oxygen (WTW, Hydrolab). Laboratory tests included analyses of pH, conductivity, alkalinity (ANC), concentration of cations and anions, concentration of total phosphorus and dissolved organic carbon. Phytoplankton (algae) and zooplankton were identified and counted in preserved samples. Concentration of chlorophyll-*a* was determined as a measure of the phytoplankton biomass.

In 1991, brook char (*Salvelinus fontinalis*) and brown trout (*Salmo trutta morpha fario*) were experimentally reintroduced to the Bedrichov reservoir and its inlets; in 1996, the Sous reservoir was stocked by brook char. An inventory of fish populations in the stream channels was done after snowmelt (May), in the summer (July) and in the period of reproduction (October) (electrofishing in transects 30-60 m long in stream channels, depth finder in reservoirs). Heavy metals in the samples of muscles and liver of fish were analysed in the laboratory. A special attention was paid to temporary and partial acidification events.

Results

Acid Atmospheric Deposition

At Jizerka, the acid atmospheric load measured in the open field culminated in the late 1980s. Sulphate was 45% and nitrate 16% of the total atmospheric deposition. In the early 1990s, the acid atmospheric load of the open field already dropped to 40% of the 1987 level. This corresponds with the decreased SO_2 concentration in the air (Figure 1) resulting from the reduced productivity of coal power stations in central Europe.

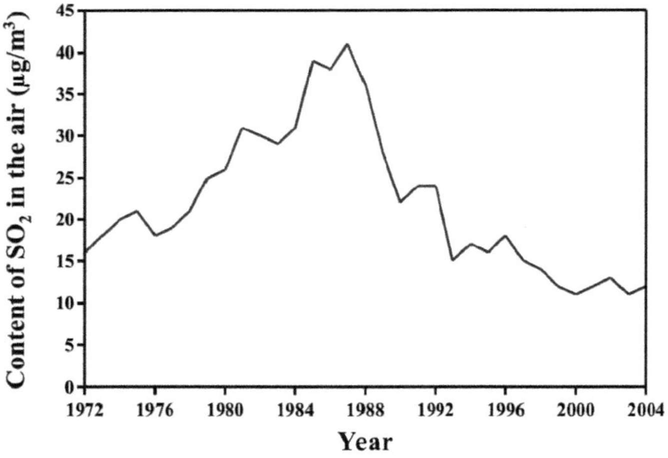

Figure 1: Mean annual content of SO_2 in the air: the Jizerka station (elevation of 860 m), 1972-2004.

Table 2: Annual deposition of sulphur in mature spruce stands and herbaceous vegetation: Jizerka (980 m a.s.l.), 2000-2004

	Spruce stand	*Spruce stand*	*Grass*
Defoliation (%)	35	75	-
Gross precipitation (mm)	1254	1254	1254
Throughfall (mm)	878	994	1056
Interception (mm)	376	260	198
Precipitation pH	4.4	4.4	4.4
Throughfall pH	3.8	4.1	4.5
Open field sulphur deposition (kg/ha)	11.8	11.8	11.8
Canopy sulphur deposition (kg/ha)	34.5	28.7	12.4

Erosion Patterns

The Universal Soil Loss Equation (USLE), Wischmeier and Smith (1978), was applied to predict the soil erosion at the Jizerka experimental basin (instrumented since 1982): the rainfall factor $R = 53$ (annual precipitation of 1400 mm), the factor of soil erodibility $K_1 = 0.22$ (sandy loam with an original infiltration capacity of 150 mm/hour), $K_2 = 0.26$ (after reduced infiltration capacity to 40 mm/hour by the harvest of timber), the length factor $L = 8.2$ (length of the slope 1000 m), the slope factor $S = 3.0$ (slope of 18 percent), and the factor of vegetation and management $C_1 = 0.01$ (well stocked woodland managed), $C_2 = 0.05$ (grass). Thus the potential annual soil loss at the Jizerka basin is $G_1 = 2.9$ t/ha (0.2 mm) and $G_2 = 17.0$ t/ha (1.2 mm).

From field observations, negligible sheet erosion occurred in run-off plots of both forest (mature spruce stands) and clear-cut (invasive grass). However, the significant loss of soil was related to the length of erosion rills produced by the harvest of timber (Table 3).

Table 3: Soil erosion in the Jizerka catchment (1982–2003)

Period	*1982*	*1982-85*	*1985-88*	*1988-92*	*1992-95*	*1995-02*
Soil loss (mm/year)	0.01	0.6	1.2	0.3	0.1	0.05
Length of rills (km/km^2)	1.3	2.4	3.9	4.7	4.7	4.7
Clear-cut area (%)	0	38	40	21	1	0
Clear-cut total (%)	0	38	78	99	100	100

From an inventory of erosion rills, 14 rills were identified in the Jizerka catchment, and 22 in the adjacent sub-basin of the Sous reservoir. The maximum depth of rills varies from 0.2 to 1.2 m related to their length and frequency of skidding. While the length evidently affects the depth of rills (Spearman R_s= 0.88, $R_{sc,\ 0.05}$= 0.45) correlation between the depth and slope is not significant (R_s= 0.24, $R_{sc,\ 0.05}$= 0.45, Figure 2).

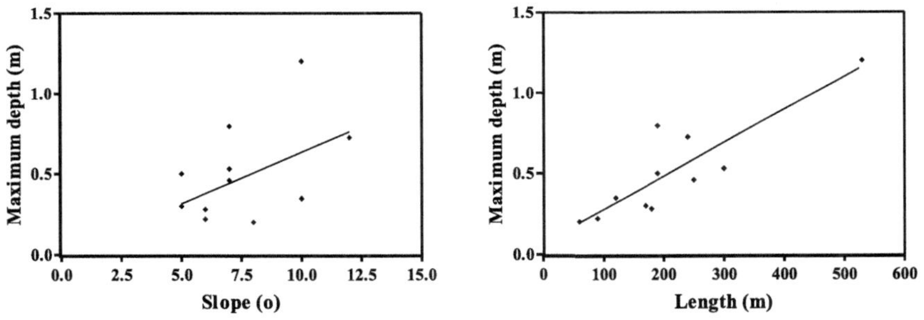

Figure 2: Erosion rills in the Jizerka catchment: maximum depth, slope and length.

Stream-Water Chemistry

In the 1970-1980s, the water quality in water courses and reservoirs of the Jizera Mts declined significantly: pH values dropped to 4-5, the content of aluminium increased to 1-2 mg/l (with a high level of toxic forms of aluminium–free Al^{3+} as well as inorganic complexes of Al), benthic fauna was reduced. In three headwater reservoirs (Bedrichov, Sous and Josefuv Dul), the extinction of fish was documented, and both zooplankton and phytoplankton were drastically reduced. Seasonal changes in the water chemistry (episodical acidification after snowmelt or rain-storm) are relatively high.

In surface waters, the first signs of a recovery were observed already in the early 1990s as a consequence of reduction in the leaf area (and the atmospheric deposition under the canopy) of spruce plantations.

After the clear-cut of spruce stands, significant changes in stream-water chemistry of the Jizerka catchment were found (Figures 3 and 4): mean annual values of pH increased from 4.0 to 5.3, concentration of sulphate decreased (from 13.0 to 6.0 mg/l) as well as nitrate (from 6.0 to 4.0 mg/l).

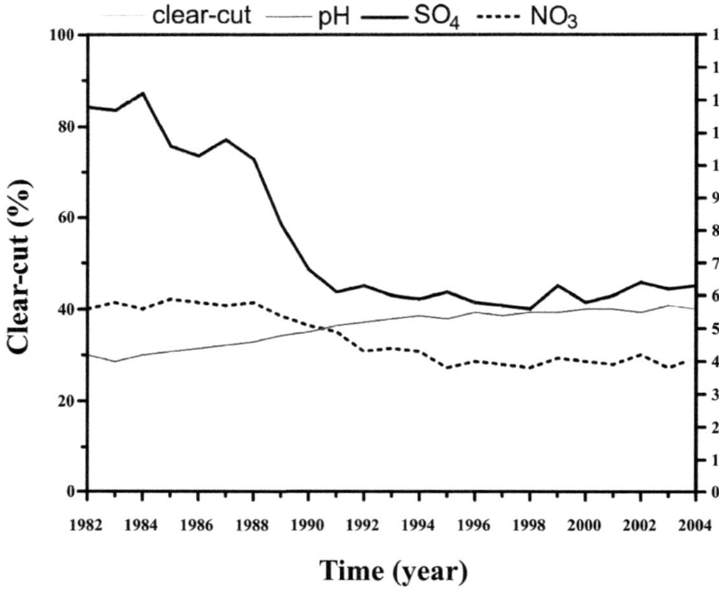

Figure 3: Stream-water chemistry and clear-cut of mature spruce stands: the Jizerka experimental basin, 1982-2004.

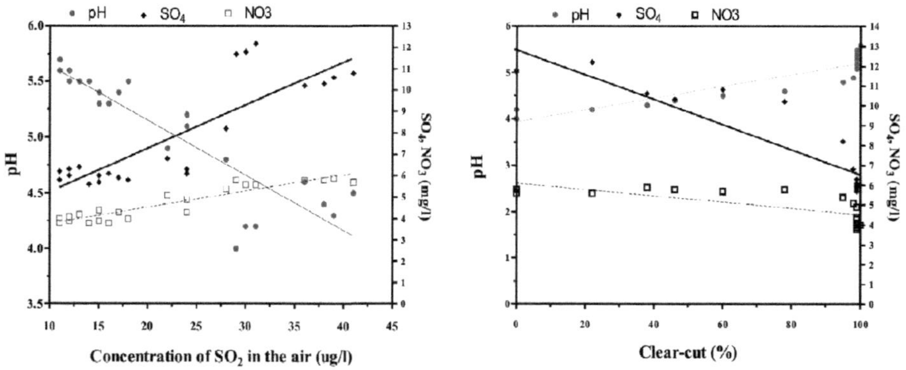

Figure 4: The effects of drop in SO_2 emissions and clear-cut on stream-water chemistry (Jizerka catchment): regression analysis (considering the effect of clear-cut on the drop of sulphate $r^2_{0.05} = 0.86$; for the effect of reduced SO_2 emissions on the drop of sulphate $r^2_{0.05} = 0.72$).

Reintroduction of Fish

Native fish species in waters of the upper plateau of the mountains were brown trout (*Salmo trutta*) and brook minnow (*Phoxinus phoxinus*). Anyway, already in the 1920s, a population of brook trout (brook charr, *Salvelinus fontinalis*) was successfully introduced into the Bedrichov and Sous reservoirs.

Since the 1950s, surface waters of the upper plain of the Jizera Mts were fishless as a result of the extremely acidified environment and the related high content of toxic metals. First signs of a recovery from acidification were observed in both the water chemistry and biota at the beginning of the 1990s. Three reservoirs were successfully re-stocked with brook trout in 1991 (Bedrichov), 1996 (Sous), and 1994/1998 (Josefuv Dul), and its populations studied for ecology, ethology, and feeding behaviour. This North American salmonid is known for its tolerance of low pH of the water but in the Jizera Mts, it shows an adaptation to extreme conditions (low temperature of the water, low pH with peaks of acidity in snowmelt and flood flows). The survival of its populations is enabled by the occurrence in a lake-tributaries system where the charr can feed, hide, mate, and find a place/water layer with a more suitable pH. The individuals reach the age of 3+ years (most females are fertile in the age of 2+), and their average body length is 104 mm in the first year of life, 155 mm in the second, and 201 mm in the third year in Bedřichov; in the limed Souš reservoir, it is 124 mm, 194 mm, and 266 mm. The brook char is a generalist feeding on the bottom (larvae and pupae of insects) and from the surface (terrestrial insects), and the fish feeds during the winter time.

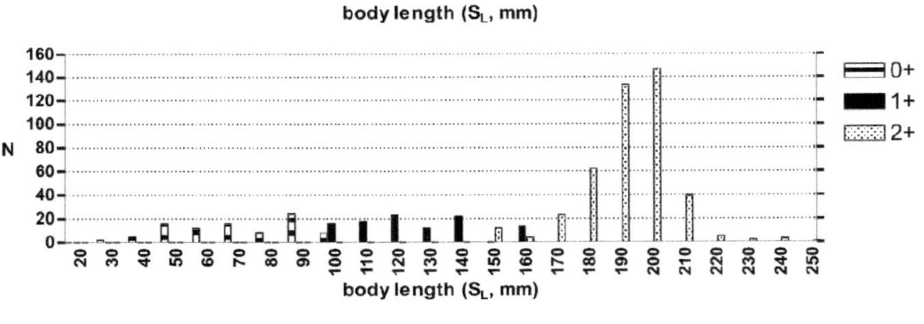

Figure 5: Brook trout (*Salvelinus fontinalis*) in the main inflow of the Bedrichov reservoir), 1995-1998: Age structure of the population, and the length distribution of individuals in age categories (0+, 1+, 2+, 3+ years).

Nowadays, the reproducing charr populations are stabilized in all the three reservoirs (Bedrichov, Sous and Josefuv Dul). However, they are still endangered by episodic acidification events with elevated concentrations of the toxic fraction of aluminium.

An experimental reintroduction of brown trout and minnow to the reservoirs was not successful but on sites with higher values of pH (in lower parts of some streams), the population of brown trout becomes successful in competition with charr.

However, the concentration of aluminium and heavy metals in the fish still exceeds the hygienic standard (Table 4). The high content of Al and heavy metals in fish tissues originates from the dominant component of food–benthic Ephemeroptera (mayflies) and Trichoptera (caddisflies, *Hydropsyche* sp. dominating)–in which extremely high values were found: 150-218 mg/kg of aluminium, 0.1-0.5 mg/kg of mercury, 0.3-4.0 mg/kg of cadmium, and 1-54 mg/kg of lead.

Table 4: Pollutants in brook char of the Bedrichov and Sous reservoirs, 1996-2000

Pollutants	*Content (mg/kg)*		*Hygienic limit*
	Muscles	*liver*	*(mg/kg)*
Mercury	0.04-2.6	0.05-6.2	0.01
Cadmium	0.1-0.4	0.4-6.2	0.05
Lead	1.4-2.7	1.3-7.0	1.0
Aluminium	6.6-18.2	13.1-90.7	30.0

Conclusions

It is evident that coniferous forests dominating in many mountain regions are not in the steady state phase (Teller et al., 1992). In the Jizera Mts, forests belong to the most sensitive ecosystems in Europe: slow weathering bedrock and shallow podzolic soils with a very shallow pool of basic cations have a small buffering capacity in comparison with the actual deposition. In the 1980s, watersheds in the Jizera Mts were stressed by an extreme acid atmospheric deposition, die-back of spruce plantations, and commercial forestry practices.

The contemporary improvement of surface water quality in the Jizera Mts seems to be a consequence of both the decreased level of air pollution, and reduction of the leaf area (and roughness of the canopy) by the clear-cut of spruce stands, and partially also by liming of both reservoirs and watersheds. The values of pH of the water increased from 4-5 to 5-6, the aluminium concentrations dropped from 1-2 to 0.2-0.5 mg/l and the brook char (*Salvelinus fontinalis*) was successfully reintroduced in headwater reservoirs.

In a long-term perspective, the water quality might be improved by planting stands near the native composition (deciduous or mixed forests with lower leaf area and surface roughness in comparison with spruce plantations). Natural deep-rooting tree species (beech, fir) can also improve the ecosystem stability: to include subsoil horizons in the cycling of nutrients, to minimise leaching of nutrients and to support stabilisation of slopes (including wind-cast susceptibility). The management of mountain watersheds should include traditional environmental friendly forestry practices (clear-cut limits, planting in advance, skidding the timber by horses or cables, seasonal skidding and manual labour re-afforestation). Respecting the riparian zones is essential for the stabilisation of mountain watersheds.

Acknowledgements

This research has been supported by the Earthwatch Institute (Boston, USA, 1991-2003: Project "Mountain Waters of Bohemia") and by the Czech Grant Agency (Prague, Czech Republic, 2001-2003: project 103/01/0322).

References

Brocksen, R.W. and Wisniewski, J. (Eds.), 1988. Restoration of Aquatic and Terrestrial Systems. Kluwer Academic Publishers, Dordrecht.

Haigh, M.J. and Křeček, J., 1991. Headwater management: Problems and policies. *Land Use Policy,* **8:** 191-205.

Haigh, M.J. and Křeček, J. (Eds.), 2000. Environmental Reconstruction in Headwater Areas. Kluwer Academic Publishers, Dordrecht.

Křeček, J. and Hořická, Z., 2001. Degradation and recovery of mountain watersheds in the Jizera Mountains, Czech Republic. *Unasylva,* **37:** 247-259.

Stuchlík, E., Hořická, Z., Prchalová, M., Křeček, J. and Barica, J., 1997. Hydrobiological investigations of three acidified reservoirs in the Jizera Mountains, the Czech Republic, during the summer stratification. *Can. Tech. Rep. Fish. Aquat. Sci.,* **2155:** 56-64.

Teller, A., Mathy, P. and Jeffers, J.N.R. (Eds.), 1992. Response of Forest Ecosystems to Environmental Changes. Elsevier Science Publishers, London.

Wischmeier, W.H. and Smith, D.D., 1978. Predicting Rainfall-Erosion Losses – A Guide to Conservation Farming. USDA Handbook, No. 537.

SECTION D

Monitoring and Mitigation of Disasters

18

Disasters in the Hindu Kush Himalayan Region: A Case Study of Tsatichhu Lake in Bhutan

Mandira S. Shrestha and Karma Chhophel[1]
International Centre for Integrated Mountain Development
Kathmandu, Nepal
[1]Hydromet Services Division, Department of Energy
Thimpu, Bhutan

Introduction

The Hindu Kush-Himalayas (HKH) is one of the youngest mountain regions and is tectonically very active; making them vulnerable to hazards that often has disastrous consequences. Intense seasonal precipitation, in the central and eastern Himalayas particularly during the monsoon months (June-September) and in the western Himalayas and the Hindu Kush during winter, triggers various types of natural hazards in different elevation zones. Snow avalanches and glacial lake outburst floods predominate at very high elevations (>3500 metres), while landslides, debris flows, and flash floods are common in the middle mountains (500-3500 m).

Almost all of the rivers that originate from the Himalayas flow through narrow gorges with very steep slopes as well as sharp bends. The occurrence of flash floods particularly in narrow river valleys is one of the most feared consequences of cloudbursts, landslides or glacial lake outbursts. There are large mass movements, leading to rock and debris avalanches deposited within the steep mountain rivers causing blockage of flow. The water is impounded temporarily behind the dam created and inundates the upstream area. Investigations have shown that 40% of the cases correspond to sliding processes considering the rotational and translational movements, 17% by rock avalanches and about 14% by debris flow. The triggering causes of such landslide dams have been found to be a direct consequence of rainfall events both short duration high intensity storms and prolonged precipitation. Most of the landslide dam failures are found to be due to overtopping or piping (Schuster and Costa, 1988). Therefore the longevity of landslide dam ranges from several hours to hundreds of years but most last several days to a few months. The occurrence of a dam failure by overtopping or piping occur when (a) the hydrological balance in the watershed area upstream from the dammed section allows the lake to fill up over and above its capacity and (b) the loss due to both seepage through the debris dam and the evapotranspiration from the lake surface is lower than the lake inflow discharge (Ermie and Casagli, 2002). A failure of such dams leads to sudden release of water downstream impacting the economy of the region and the safety of the communities and creating loss of life and property.

In 1883, a largest known landslide dam in the Central Himalayas blocked the Birahiganga in India to form a large dam of 350 m height. The lake formed behind the dam was breached in June 1970 releasing a large volume of water down the valley raising the level of river by 50 m at Srinagar, a town almost 110 km downstream. Similarly on 8 July 1993, a rockfall in the Naptha area blocked the flow of Sutlej River for about 30 minutes, creating a 6 km long temporary lake. The Sanjay Vidyut Pariyojana was damaged due to the water entering the hydroelectric installation which resulted in the plant shutdown for three long months and a loss of millions of dollar revenue (Sah and Virdi, 1996). Recently just upstream in the Pareechu River a tributary of Sutlej River in Tibet, China, there was a landslide in June 2004 which formed a natural dam holding about six million m^3 of water by July 2004. On 8 July, another landslide occurred and blocked the river at a location about 30 km from the China, India border, forming a natural dam of about 35 m. The total water volume increased to about 79 million m^3 by August due to continuous heavy rain (CMA, 2004). Villagers were evacuated and people were made aware of the imminent disaster should the dam fail. The landslide dam however did not fail and the water is still impounded behind the dam.

In April 2000, about 300 million cubic metres of debris, soil and ice dammed the Yigonzangba River—a tributary of the Yarlanzambo also known as the Bramhaputra River. This natural landslide dam was about 130 m thick, 1.5 km wide and 2.6 km long, created in only eight minutes. The Chinese Government spent US$ 7.2 million to construct a spillway through the dam (ICIMOD, 2001). Despite mitigative measures taken by the Chinese Government, the dam burst on 10 June 2000 and 30 people were dead with more than 1000 missing in Arunachal Pradesh of India resulting in an economic loss of US $ 22.9 million.

Bhutan is a mountainous country on the eastern fringe of the Himalayas. It is bounded on the north by the Tibetan Autonomous Region of the Peoples' Republic of China and on the south by the Indian states of Assam, West Bengal and Sikkim. The elevation ranges from 100 masl at the foothills to more than 7500 masl at the border with China. The Glacial lake outburst flood in October 1994 from the Lugge Tsho located in eastern Lunana area caused extensive damage both to lives and property. Twelve houses were damaged, five water mills were totally washed away, and 816 acres of dry land was damaged (Mool et al., 2001). Similarly there has been occurrences of several landslides due to intense monsoon rain and formation of lakes. This paper describes in brief the Tsatichhu landslide and damburst process and the timely mitigative measures that were taken to enable the prevention of loss of lives and valuable property downstream.

Tsatichhu Landslide

Location of Tsatichhu Landslide and Downstream Area

The Tsatichhu landslide occurred in the Jarey Gewog, Lhuentse district, East Bhutan on September 10, 2003 due to a slope failure. The Tsatichhu river is a tributary of the Wabrachhu the confluence of which is located about 1.5 km from the lake outlet. Ladrong, a small village with about 50 people, is situated opposite the lake near the confluence. The Wabrachhu is a tributary of the Kurichhu the confluence of which is located about 6.8 km downstream of the confluence of Tsatichhu and Wabrachhu. A small settlement Austsho is located at this confluence. The whole 6.8 km of channel to Austho is narrow, confined and steep with an average gradient of 6.5°. From the confluence of the Kurichu and Wabranchhu, the Kurizampa and KHP are located at 21 and 25 km respectively. The location is shown in Figure 2. There are no major settlements along the way. However the Kurizampa bridge is a vital link in the east-west highway connection. There is a 55 m high dam, and a powerhouse at the toe of the dam with a switchyard between the dam and powerhouse. The Kurichhu powerhouse is made available to the surrounding districts in Bhutan and the bulk is exported to India. Hydroelectric export is a significant source of revenue for Bhutan and currently about 45% of the national revenue is from hydropower. In the coming year from July 2006 after the commissioning of the Tala Hydropower project it is expected to contribute 75% of the national revenue. As per the eyewitness record of villagers of Ladrong village the landslide occurred just after midnight and was accompanied by ground shaking and rumbling noise. The lake has formed in a remote area where access is through tough and difficult terrain.

Hydrology of the Basin and Landslide Area

The catchment area above the Tsatichhu lake derived from a 1:50,000 topographic map is 24.3 km^2. The area receives an annual precipitation of about 1500 mm. Just before the occurrence of the landslide the area received heavy rainfall higher than the normal. In December the water level of the lake was maintained at 1646 m with a limited amount of overflow. The amount of inflow into the lake was estimated to be about 0.5 m^3/sec. There was some overflow from the dam including the seepage of water from the landslide scar.

Figure 1: General location of Bhutan.

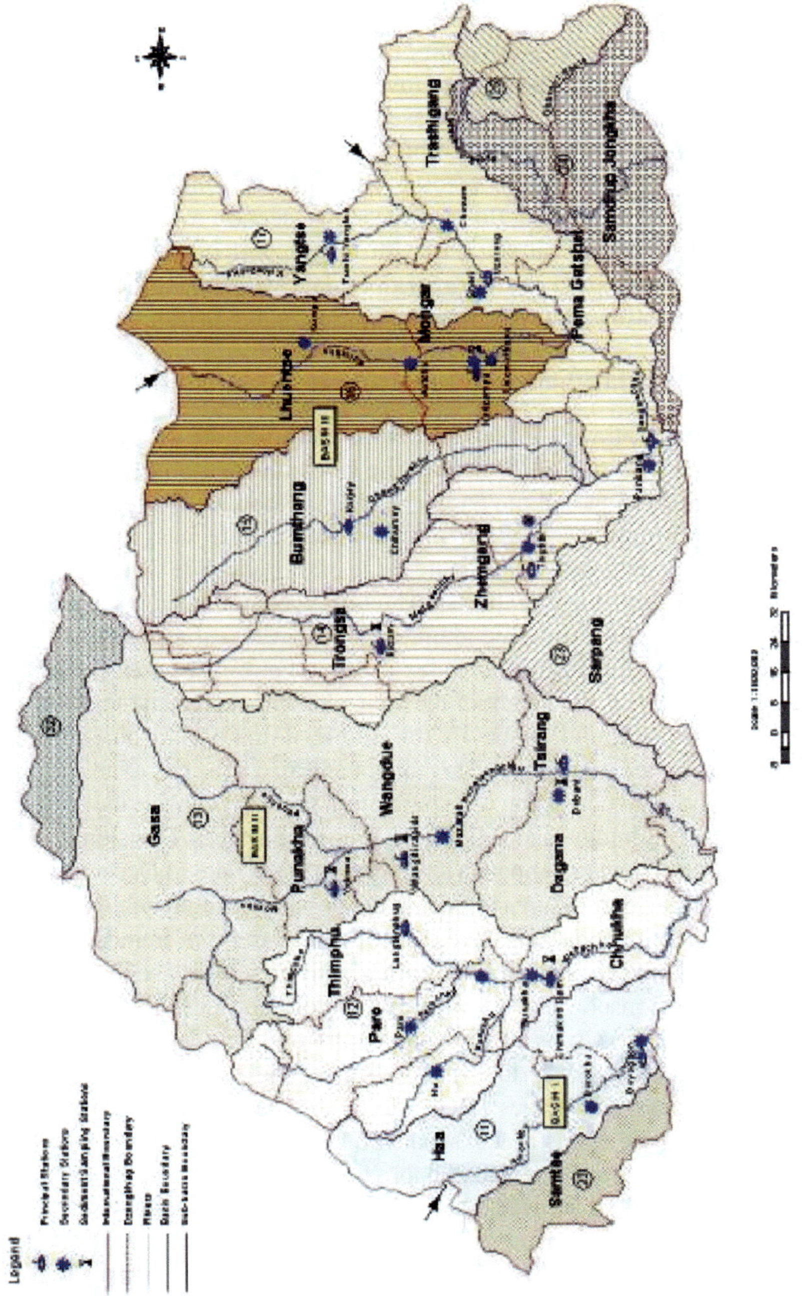

Figure 2: Location of Tsatichhu lake with respect to the Kurichhu/hydropower project.

The catchment area above the Kurizampa is 8600 km^2. It is estimated that about 35 percent of the catchment area lies in Tibet, China. There is a gauging station at Kurizampa with a cable way. Long-term daily flow data is available at the station from the year 1991.

The Hydromet Services Division of the Department of Energy, Ministry of Trade and Industry of the Royal Government of Bhutan is the mandated organization to collect, analyze and disseminate the hydrometeorological data and information. It coordinates with other organizations such as the Department of Agriculture, Department of Geology and Mines. Under the Hydromet services division there is a flood forecasting section which is currently monitoring 10 hydrological stations with the support of the Central Water Commission of India and sharing real time data with India under a mutual agreement. One of the stations is on the Kurichhu at Kurizampa.

Mechanism of Landslide and Composition of Landslide Dam

The landslide mass travelled down the slope and filled the valley floor of the Tsatichhu river. The size of the scar of the landslide measured to be 0.6 km^2 with the top at 2345 m and the base at 1710 m (Noord and Gyenden, 2004). An estimated volume of 33 million m^3 debris of the slide mass formed a very sizable landslide dam with an estimated size of the landslide dam body of 0.3 km^2. A random assessment of the Tsatichu lake was conducted by an expert in October 2003. The landslide dam stability was estimated using the geomorphological dimensionless index using the physical parameters of the landslide dam such as the height of the dam, catchment area and the landslide volume and was found to be in the stable domain (Norbu, 2003). A multidisciplinary team was formed to monitor and assess the risk of an imminent damburst. The team consisting of experts from various agencies including the Department of Geology and Mines, Hydromet Services Division and the Kurichhu Hydro Power Company (KHPC) visited the dam in December 2003 to explore mitigative measures including the possibility of breaching the dam. Preliminary recommendations were that the dam was not of immediate threat of breaching. However intense monitoring activities were needed to ensure that during the monsoon the risk of breaching did not increase.

The field observations of the topography and the composition of the landslide conducted during the visit indicated that the landslide probably happened in successive stages (DGM, 2003). At least two stages could be distinguished on the dam surface, separated by a slight depression which ran parallel to the valley at a distance of approximately 150 m from the left valley side. The landslide scar showed that there are two different phenomena. In the southern part there was a planar rock slide which contained a large volume of rocks. In the Northern part the mass movement had only involved the overburden soil cover and vegetation without appreciable amounts of bedrock and the presence of the original vegetations at some locations on the slope. Due to the complex development of the landslide dam it was very difficult to predict the spatial distribution of the different types of materials inside the landslide dam. From the observations of the landslide source area it was estimated that about 70 to 80% of the material comprised coarser fractions while the remaining percentage was gravel, sandy material and fines generated due to impacts from the original burden. The surface of the landslide was built up of large angular boulders of granite-gneiss and augen-gneiss up to 5 m in diameter.

Monitoring and Mitigation Activities to Prevent a Likely Disaster

Several field visits were made to assess the stability of the lake and preventive measures that need to be taken to minimize the loss of lives and property downstream. The multisectoral team met on a regular basis to monitor the developments of the lake and the potential risk of dam failure as well as the execution of the assigned tasks. The Department of Energy had the responsibility of the overall coordination. The Hydrology section of the Hydromet Services Division installed a water level recorder to monitor the level of the lake and to calculate the inflow. The Department of Geology and Mines carried out a downstream hazard assessment. At the recommendation of the multisectoral team a spillway of 50 m length and approximately 4 m wide was excavated by beginning of May to facilitate easy overflow along the channel. On 21st May a large slide occurred and an estimated eight million cubic metres of mud and stone debris was washed away. The debris was carried downstream which created a second lake on the Wabranchhu the tributary of the Kurichhu. An inspection of the site was made by a team of experts on 24th and 29th May. The status of the lake as of 24th May was reported as:

Height from sea level upto lake water level	- 1645.50 m
Depth of lake water	- 115.50 m
Length of the lake	- 1.45 km
Width of the lake	- 300.00 m
Volume of the lake	- 12.25 million m^3
Seepage from dam body	- 0.57 m^3/sec
Overflow from spillway channel	- 2.0 m^3/sec

The discharge into the lake was measured to be 0.5-0.6 m^3/sec. Due to the slide the thinning of the dam top had occurred. The dam crest was about 200 m long with the base about 600 m. The thickest side was about 103 m at the right side, 70 m at the centre and 55 m in the left side. The total volume of water stored behind the dam was estimated to be 12.5 million m^3 and in the event of a burst the peak discharge was estimated to be about 2500 m^3/sec lasting for about two hours. A view of the landslide dam and the surrounding taken on May 29th is shown in Figure 3. It was recommended that an early warning system be put in place to minimize the impact on human and property downstream in the event of the dam failure.

Figure 3: A view of the Tsatichhu landslide dam on 29th May 2004.

Installation of Early Warning System

To minimize the loss of lives and property—in particular to warn the KHP located 25 km downstream—a need to install an early warning system was imperative. For data communication HF radio sets were installed at two locations, one at Ladrong and the other at Autsho. Kurizampa and the dam control room at KHPC were already equipped with HF radio sets. Senao telephone sets were also provided with four spare batteries. Observers were kept on high alert 24 hours and instructed to inform KHPC on any increase in seepage through the dam or rapid increase in water level of the river. The villagers of Ladrong, Autsho and KHPC were alerted. Arrangements were also made to provide early warning to downstream in India in Barpeta directly through HF radios from Ladrong by the Flood Warning Section of the Hydromet Services Division.

Tsatichhu Dam Burst

Mechanism of Burst

The triggering mechanism of a landslide dam failure is similar to that of a GLOF. Both the events depend on the nature of the damming materials, the position of the lake, the volume of stored water, physical and topographical conditions and other physical conditions of the surroundings. On July 10th, a major part of the Tsatichhu dam failed at around 16:30 hrs due to sustained monsoon rainfall and additional erosion in the flow path in the first week of

July and a large volume of lake water was released downstream. The breach profile was V-shaped on the extreme left of the dam where approximately 40 m of material was removed with a breach width of about 100 m as shown in Figure 4.

Figure 4: Tsatichhu Lake looking downstream.

The release of high discharge from the Tsatichhu Lake travelled downstream creating a short blockage at Autsho but was removed by the high discharge of the Kurichhu. Some stretches of low lying road sections south of Autsho and close to Rewan were washed away. The observers who were kept on high alert provided timely warning to the KHPC of the landslide dam failure. The first waves of the flood reached KHP at about 18:30 hrs. The flood wave took approximately two hours to travel to Kurizampa. Due to the timely warning the spillway gates at the damsite were raised to enable lowering of the reservoir level; hence causing very little damage to the KPH infrastructure.

Downstream Assessment of Impacts

The large volume of water including tonnes of logs and debris had passed through the dam by 20:00 hrs. The flood had deposited a lot of silt and logs in the tailrace area of the power house (Kuensol, 2004a). The flood washed away a transmission tower about 15 km downstream of the KHP. Other transmission towers were also bent. Without the towers and the line, the KHP was desynchronized from the Indian Grid (Kuensol, 2004b). The KHP was shut down for about a month. As per the recommendations of the multisectoral team a downstream hazard assessment was carried out based on the data collected during the fieldwork in May 2004. The assessment included hazard identification, scenario development, vulnerability assessment, impact analysis and hazard preparedness and awareness raising. The report of the hazard assessment was however only completed after the occurrence of the dam failure. The likelihood scenario analysis presented in the report was very close to the actual event.

Hazard that Remains at the Landslide Dam Area

To assess the hazard that remains, the Hydromet Services Division made a visit to the Tsatichhu and Wabrachu lake areas in December 2004. The sudden and instant lowering of the lake level by 40 m after the dam breach triggered secondary instability. The support provided by the lake water was abruptly removed due to lowering the lake water level which induced slope failure in the saturated soil slopes. The large volume of loose or unconsolidated material (landslide debris) that is present in the remaining dam body, along the Tsatichhu river bed and along the Wabrachu also retains a certain degree of hazard. The dam body of the Tsatichhu landslide is not yet stabilized. There is also continuing seepage from the landslide dam. During the field visit a bathymetric survey was conducted of both lakes to determine the volume of water in the lakes. The breached outlets were widened to facilitate the flow of larger discharges and minimize the water collection in the lake to reduce the hazard. The debris flow originating from the Tsatichhu landslide dam has altered the whole morphology of the Tsatichhu valley. Large volumes of debris have been removed and deposited in the Tsatichhu valley. This layer is estimated to be about 20 to 30 m thick and at the confluence with Wabrachu is about 15 m. The confluence area of Wabrachu and Kurichu, just downstream of Autsho, was temporarily filled with debris flow deposits of Wabrachu. During the field visit there still remained a fresh deposit of about 4 m on top of the existing alluvial fan. The flood warning section will continue to monitor the water level of the two lakes and the seepage as well as the downstream areas for another monsoon season. The Tsatichhu lake after the landslide dam failure is shown in Figure 5.

Lessons Learnt and Recommendations

The joint effort of the mitigation activities to reduce the risk of Tsatichhu landslide dam failure has been unique and well coordinated. This landslide dam formation and subsequent failure have contributed to better understanding of the landslide dams and lakes and the natural hazard that they pose in mountain areas. It has also shown the importance of early warning system and the need for a coordinated approach to enable timely warning downstream to reduce the loss of life and property. The comparison of the actual event with that of the hazard assessment showed a very close agreement indicating the importance of hazard assessment. Hence in the case of any future landslide

Figure 5: The Tsatichhu Lake after the landslide dam failure.

dammed lake occurrence it is highly recommended to perform a comparable hazard assessment. After the occurrence of the landslide dam failure it is important to evaluate the conditions of the site to assess the hazard that still remains since remaining large volume of material in the future could cause problems of blockade as well as occurrence of large debris flow. Thus in mountain areas awareness building, environmental hazard mapping, continuous monitoring and installation of early warning systems are essential in disaster reduction.

References

Ermini, L. and Casagli, N., 2002. Prediction of the behaviour of landslide dams using a geomorphological dimensionless index. *Earth Surface Processes and Landforms,* **28:** 31-47.

CMA, 2004. Imminent Damburst in Tibet. International Cooperation Department, China Meteorological Administration (CMA), Beijing China. Personal communication, 13 August 2004.

Mool, P.K., Wangda, D., Bajracharya, S.R., Kunzang, K., Gurung, D.R. and Joshi, S.P., 2001. *Inventory of Glaciers, Glacial Lakes and Glacial Lake Outburst Floods; Monitoring and Early Warning Systems in the Hindu Kush-Himalayan Region.* United Nations Environment Programme, Regional Resource Centre – Asia and the Pacific (UNEP/RRC-AP), Bhutan and the International Centre for Integrated Mountain Development, Kathmandu.

Noord, H.V. and Gyenden, L., 2004. The Tsatichhu Landslide Dammed Lake, Slope Stability and Downstream Hazard Assessment and Some Reflections on the Lake Outburst. Department of Geology and Mines Report, Royal Government of Bhutan, Thimpu, Bhutan.

Norbu, P., 2003. A Random Assessment of the Tsatichhu Lake with Sideways Look at Possible Behaviour of the Landslide Dam that Caused the Lake Formation, Thimpu, Bhutan.

DGM, 2004. Stability Assessment of the Tsatichhu Dam and Lake. Department of Geology and Mines (DGM), Royal Government of Bhutan, Thimpu, Bhutan.

Kuensol, 2004. Tsatichhu Lake is not a Threat. Kuensol Report, June 5, 2004.

Kuensol, 2004. Tsatichhu Lake Bursts, Kurichhu Shut Down, Kuensol Report, July 2004.

Sah, M.P. and Virdi, N.S., 1995. Channel blockade and flash floods: Some examples from the Beas and Sutlej Valleys of Himachal Pradesh. Paper presented at the Interactional Meet on Natural Hazards in Himachal Pradesh and Their Mitigation, October 14-15, 1996, Shimla, Himachal Pradesh, India.

ICIMOD, 2001. Mountain Flash Floods. *ICIMOD Newsletter,* **38,** International Centre for Integrated Mountain Development, Kathmandu, Nepal.

19

Perception and Communication of Flood Risk: Preliminary Results from the FLOWS Project

Irina Krasovskaia, Anja Skiple Ibrekk, Lars Gottschalk[1] and Hallvard Berg

Norwegian Water Resources and Energy Directorate, Oslo, Norway

[1]Department of Geosciences, University of Oslo, Oslo, Norway

Introduction

Floods cause enormous economic damage, as well as loss of life, and the tendency is increasing throughout Europe as evident from many sources (e.g. EM-DAT, 2004). Floods represent, however, a natural phase of river flow regimes and a flood hazard cannot be eliminated. It is the vulnerability of the society that should be assessed instead. In the last two decades "resilience" has become the buzzword (World Disasters Report, 2004). As it is not possible to provide total flood safety using even most advanced technical measures, it is of vital importance to learn how to live with floods by means of better preparedness; better forecasts; better spatial planning; better perception of flood hazard; and retrofitting.

A lot of effort has been put on modelling, forecasting and mitigation measures to reduce the damages of floods, while "non-structural measures" have yet been given much less attention. "Governance" is a new catchword to highlight the importance of the soft components of water resources management (Falkenmark, 2004). Though many perception studies in connection to floods have been performed during the last decade (e.g. Krasovskaia et al., 1995; Morris-Oswald and Simonovic, 1997; Horven Skellnes, 2001; Environment Agency, 2004), in most cases no attempt has been made to incorporate this information directly into spatial planning practices.

The main objective of the Interreg III B FLOWS Project (Flood plain Land use Optimising Workable Sustainability) is to improve all aspects of flood management with special emphasis on spatial planning. The Project period was 2003-2006 and the participating countries were Germany, the Netherlands, Norway, Sweden and the UK. The Project involves local, regional and national authorities, as well as several universities. Project leader is Cambridge County Council and the Norwegian Water Resources and Energy Directorate (NVE) is the lead partner in Norway, coordinating the part of the Project studying similarities and differences in the perception of risk among the public and decision-makers as well as demonstrating various information dissemination techniques in all the five participating countries of the North Sea region.

This article summarises the main findings of a number of project reports in the frame of the FLOWS Project focusing on the social aspects of flood risk like its perception and the communication.

Why Study Perception of Flood Risk?

Floods represent a threat only with respect to human society. In a democratic society a risk assessment policy should be based on the agreed criteria for tolerable ("acceptable") risk between the individuals and the society (Krasovskaia et al., 2001). To reach a consensus about tolerable risk an open dialogue between public and decision-makers is indispensable. Flood risk perception by the general public and decision-makers is essential information for such a dialogue as well as for decision making concerning all the steps in flood risk assessment from preparedness and forecasts to spatial planning and retrofitting. Communication of risk is often considered to be in the heart of all risk problems. An efficient communication between the expert and the public requires that the perspectives of both groups concerning flood hazard are known. A study of flood hazard perception by laymen and decision-makers and experts helps to highlight the gaps in understanding and communication and indicates possibilities for a better communication of flood hazard stimulating active public involvement in flood assessment.

Why Demonstrate Disseminating Techniques?

The FLOWS Project aims to improve the links between experts, decision makers and the public by investigating and demonstrating different techniques of disseminating flood information in the participating countries. Effective flood risk management depends both on the input of usable technical information from experts to decision makers and the communication of a clear message from decision makers to and from members of the public. The challenge for every country is finding ways that aid the understanding of floods and engage communities. The Project will establish best practice by considering various different dissemination techniques, including the use of GIS and the internet, signposts, symbols, warning systems, information centres and practical demonstrations, such as the retrofitting of buildings. Many activities are related to increasing awareness among school children. To improve flood assessment in general, the link between perception of flood hazard and how flood risk is communicated should be investigated and considered.

Methods: Risk Perception

Poll Study

To get insight into how the flood hazard is perceived by the general public polling was carried out by means of telephone interviews and involved totally 4,000 people living in flood prone areas in five countries viz. Germany, the Netherlands, Norway, Sweden and the UK. The main topics taken up during the polls were (Krasovskaia, 2005a): general awareness and concerns about flood hazard; previous experiences from floods; reasons for living in a flood prone area; knowledge about flood assessment in their region; preferable channels of information; confidence in the ways public authorities handle flood hazard; and willingness to take responsibility for strengthening resilience.

The poll was carried out by TNS Gallup and its partners in the participating countries. Interviews were conducted as CATI (Computer Assisted Telephone Interviews) in each country and each interview took about 10 minutes. The same questionnaire consisting of 32 questions on the focus topics and 10 questions on personal background was used in all countries. The original language was English (Master questionnaire) and the questions were translated to the local languages with some slight local adaptation of text.

Expert Panels

Experts and decision makers received questionnaires with questions similar to those posed to laymen during polls. The choice of the respondents was the responsibility of each country. The ambition was to obtain a representative sample with respect to the existing national practices in flood assessment in the context of spatial planning. The main tasks were: getting insight into the perception of flood hazard by the experts and decision makers in the partner countries; identifying similarities and differences in the answers between the experts from different countries; identifying similarities and differences in the opinions of experts and laymen about similar topics; and identifying the important topics for discussion at national and international expert panel meetings (Krasovskaia, 2005b). During these panels the opinions about flood hazard revealed during the poll study were presented to the experts and decision makers encouraging them to discuss the discrepancies between the groups in search for a consensus. At the concluding international panel which gathered experts and decision makers from five participating countries and representatives of NGOs the discussions continued symbolising an active dialogue between laymen and decision makers. Such a dialogue is a necessary premise for an efficient assessment of flood hazard. The expert panels offered an effective platform for exchanging experiences and opinions promoting national and international networks of experts and decision makers in flood management.

Focus Groups

To gain a deeper insight into the attitudes and perceptions among people interviewed during the telephone poll, qualitative studies by means of focus groups were undertaken in two partner countries. The overall aim was to elucidate laymen's views on floods in more depth— what they think and why—as a complement to describing the patterns of views on flood hazard in partner countries. Two focus groups meetings were organised in Norway and two in the UK. Each focus group, consisting of 15-25 people, had an open discussion around topics related to the flood hazard and flooding led by a professional facilitator. People usually started with talking about practical considerations, such as the structural effects of floods, to continue with feelings of stress and frustration (e.g. about difficulties to get economic compensation) and finally more emotional concerns, such as their own safety and safety of their families (Rosslyn Research, 2004; TNS Gallup, 2004).

Results: Risk Perception

Some key findings from the polls and expert panels studies (Krasovskaia, 2005a, b) are presented below. The findings from the focus groups confirmed to a large extent the results from the poll study.

Poll and Focus Groups Studies

Limited interest in flood hazard and negative attitudes towards floods

Both poll and focus group studies showed a limited interest in and underestimation of flood hazard by people living in flood prone areas. Poor knowledge and unwillingness to accept (and assess) flood hazard were in many cases the reasons for this dissonant perception. Misunderstanding of the nature of floods revealed by the answers resulted in a passive attitude towards actions to reduce losses including simple search of information about what can be done. "This cannot happen to me" attitude was overwhelming.

The majority of respondents could not see any positive effects connected to floods (even though a list with suggestions had been offered). "Floods bring only harm!" was the dominating opinion.

Such attitudes were also confirmed by other studies attempting to find explanations through psychological ("it won't happen to me") and social mechanisms (where alarmism is unpopular and therefore avoided). Flood is a hazard that tends to be viewed as natural and familiar, characteristics that may lessen the perceived seriousness of the hazard (Enander, 2005).

False feeling of safety

Only less than half of the respondents in the poll study were aware of living in an area at risk of flooding, and many people seemed to feel safe stemming from their failure to appraise flood hazard realistically. The fact that "nothing has happened so far" seems to have reassured many. Awareness about flood hazard was highest in both Scandinavian countries (six to seven out of ten respondents) and lowest in Germany (only one out of ten was aware of living in an area at risk of flooding). Another reason for feeling safe was an exaggerated belief in flood defence and unrealistic expectations of "absolute" protection. Other studies also show that people may have been lulled into a false sense of security by the long period since the last devastating flood and also by the technical protection measures undertaken (Kundzewicz, 1999).

Passiveness with respect to raising the flood safety of their homes and lack of confidence in public authorities

The majority of respondents in the poll study have not taken any steps to prepare for flooding and reduce losses caused by it. When asked if they would be willing to make further investments in their property to reduce their risk of flooding and the impact of potential flood damage, the overriding response was "no". Almost none were ready to move to safer areas because of the flood hazard staying mainly due to sentimental reasons rather than difficulties to sell their property, finding jobs etc. People clearly preferred to delegate the responsibility for flood protection to authorities but at the same time only about half of the respondents in the poll study confirmed to be very or quite confident in the way public authorities handle the flood hazard. Previous bad experience and lack of priority by the authorities were the dominant reasons for not relying on them in this matter. Participants in the focus groups attributed an increase of the flood hazard to increasing and poorly controlled new developments in flood prone areas.

Unwillingness to take preventive measures can partly be explained by the general human tendency to avoid sure losses and to ignore low probabilities. Behavioural science research shows that perception of risk is not always sufficient to motivate people to take action, nor even personal experiences of the hazard. Understanding these mechanisms is important in order to create more effective conditions for dialogue and development (Enander, 2005).

Preferred information channels

People in the survey gave preference to information sources which "serve" information rather than having to look for it themselves. Traditional information sources like newspapers, radio and TV, and leaflets proved to be the preferred information channels in all partner countries. Germans privileged websites more often than others and in general expressed high preference equally often (five in ten respondents) to all other listed information sources apart from e-mail. However, it was generally felt that the information supplied was insufficient or inadequate. British and Norwegian respondents were those most satisfied with the information provided on the flood hazard while rather many Germans, Swedes and Dutch felt being ill informed. A possible connection between climate change and increased flooding frequently discussed in mass media seems to be one of the few topics in the context of flood hazard that have captured the attention of laymen (as evident from discussions in focus groups) and also experts.

Differences between the countries

There are more similarities than differences between the countries in the way people perceive flood hazard. However, some differences were identified in the perception studies performed. Among major differences we may note a somewhat better awareness of the flood hazard by the Norwegians and much lower among the German respondents. These latter demonstrated at the same time lower tolerance for flooding of own houses and somewhat lower confidence in the ability of public authorities to handle flood issues, while among the Swedes this confidence was somewhat higher than average. While Germans quite consequently showed higher willingness to invest in flood safety, the British respondents were most reluctant to do this. The Dutch respondents showed low concerns about flood hazard and higher acceptance of environmental changes to increase flood safety than on average. There was an obvious diversity of opinions about whose responsibility flood safety was. While local authorities rather than central were given such a responsibility by the Swedes, eight out of ten Germans and seven out of ten Dutch and Norwegians considered it to be a responsibility of the central government. The British respondents supported equally often both these alternatives. In Germany about one third of the respondents considered that individuals in danger should pay, while in Scandinavian countries only six out of one hundred respondents supported this alternative.

Expert Panels Study

The study revealed many similarities but also differences in the perceptions of flood hazard by the experts in the partner countries. Among the perceptions that were very similar are:

- Obvious awareness of the flood hazard. Experts reckoned on flooding of homes in the countries of the North Sea region with the exception of the Netherlands but showed a reasonable attitude towards flooding as a natural event that might also have some positive effects (e.g. stimulating investments in preventive measures, triggering more social responsibility and drawing public attention to the communities etc.).
- In general the existing defence measures were not considered to provide an acceptable level of safety to the citizens.

- Information on flood management and defense measures was appraised as adequate and user friendly but the importance of regular information to laymen was perceived as crucial only in the UK. Traditional information links were given preference, electronic information being also appreciated.
- An obvious need to enhance consideration of flood hazard in local planning in all the countries but the UK.
- The majority of the experts attributed the increase in costs to increased costs for precautionary measures (especially the Dutch experts) but also several other costs were considered to have increased, such as costs for direct flood mitigation, information, insurance and forecasts (especially by British and German experts).
- Public participation in flood assessment has still not found its true value in the perception of the experts and decision makers with the exception of the UK. In some countries, and especially in the UK, there were indications of efforts undertaken to actively involve laymen in flood assessment.

Among the main differences noted were: a clear leadership of the UK what concerns appraisal and efforts to actively involve laymen in flood assessment; not reckoning on a major flood event in recent future nor on flooding by the majority of the Dutch experts and higher estimation of responsibility of regulation companies for flood safety; reckoning on regular flooding of homes by the German experts, who also took the leadership in the direct contacts with public; the Swedish experts were those who were most convinced that a major flood event could happen and were most unsatisfied with the existing flood protection and the information on flood matters; among the Norwegian experts that the acceptance of flooding of private homes and natural areas seems to be the lowest and they showed the lowest appreciation of the importance of public participation in flood management and defense and information on flooding.

Many of the differences noted depend on the differences in practices of flood management and defense applied in the partner countries. It indicates that the approaches are really many and calls for an active exchange between the decision makers from different countries to really benefit from the positive experiences of their colleagues and avoid repeating the same mistakes. An international expert panel offered an opportunity for such an exchange.

Promoting Active Citizenship

Comparing the views on similar topics expressed by laymen and experts it was easy to see that there existed some obvious discrepancies. For example, while the majority of laymen were not aware of the flood hazard the experts clearly were. Acceptance of flooding of homes and other property was much bigger among the experts than laymen. There was a clear discrepancy in attitudes towards floods: while for laymen they were mostly negative, the experts could identify some positive effects in connection with floods. The knowledge about flood defence measures was obviously insufficient among the population while the decision makers were very well informed on this matter. While only few decision makers considered flood protection to be adequate, laymen showed a tendency of over-belief in protection and defence measures undertaken. The majority of experts appraised information on floods as adequate but this opinion was not shared by the majority of laymen. Also appraisal of websites for public flood information appeared to be much more popular among the decision makers than laymen. Experts sometimes attributed people higher responsibility for investments in flood safety than people themselves were ready to take.

A way to assess the discrepancy in opinions about the flood hazard is through an open dialogue between the experts and decision makers and laymen. The participants in the international expert panel agreed that public involvement was needed for new projects and promoted regular contacts and quick response to public enquiries. An information system maintained by the authorities might serve as a regular link to general public provided it is easily accessible and open for enquiries. It was noted that active public participation in flood assessment requires well defined structures what concerns responsibilities, roles and possibilities. Such structures are often missing or are at a very early stage of development. An important role in maintaining the dialogue belongs to citizens' organisations. "The National Flood Forum" and "Oslo River Forum" are examples of organisations from the UK and Norway, respectively that are already taking an active part in planning processes concerning water and providing citizens with advice and information. Representatives from both organisations took an active part in the discussions during the international expert panel bringing forward the views of laymen (Dhonau, 2005; Tønsberg, 2005).

Flood Information and Communication

Three different disseminating techniques will be demonstrated within the FLOWS project by all participating countries. At present only a few have been finalised while most of the activities are still ongoing. At the end of the project in 2006, all information techniques will be evaluated based on the knowledge of the different countries' flood management structure and practices.

There are two main types of flood information: general information to raise awareness and information given during and after a flood event ("crises information"). Many examples on flood information and communication have been investigated within the FLOWS Project, both general information towards the public and information during a crisis situation. Selected activities on flood information and communication are described below.

School Projects

Various school projects have been initiated in Norway, the Netherlands and the UK. These projects have a special emphasis on activities for school children, e.g. art work development (sandstone carving), flood symbols, flood newspaper and various events. These activities help children within the local community to understand what a flood means and accepting it as a natural phase of river flow regimes. It was considered to be important to teach about floods together with environmental issues and water management in general. The school projects got much attention in local and regional mass media and thereby propagated flood information to many more people than the school children and families directly involved.

Information Campaigns and Desks

Various flood awareness campaigns performed may serve as examples of information to the general public, like for example mobile information points placed on a community vehicle. Different types of information display related to retrofit of a heritage building are under development by a university. Practical information about what residents can do themselves before a flood occurs have also been distributed together with the responsible authorities. Leaflets describing what can be done when a flood occurs are also under development by the local municipalities.

Local Flood Warning

Improvement of local flood warning system is another activity in the FLOWS Project. In some areas, the flood situation develops very quickly and sometimes the flood is declining already when the national warnings reach the local community. In one municipality, water level transmitters have been installed sending an alarm to a mobile phone of the local duty watch who coordinates the flood warning with national authorities. The flood warning is then distributed to the police and the inhabitants living in flood prone areas by transmitting collectively by UMS (Unified messaging system) to registered stationary or mobile phones in the actual area.

Overview of Existing Flood Information

A description of possible ways to collect information is another example that will be useful to experts and decision makers. In many areas flood hazard maps are not yet available for local planners and decision makers. In such cases historical information may be useful (e.g. information on historical flood events, water level marks). Such information is found more and more often on internet. There is a need to collect all available information in databases assuring easy access. A prototype of a flood database will be assessed as part of the FLOWS Project.

Visualising Water Level

"Flood columns" is a new concept of visualising flood risk in residence areas that has been developed in the frame of the FLOWS Project. A set of vertical

Plexiglas pipes is set up in front of flood prone buildings and filled with water up to the predicted flood level. These columns are meant to give residents a better understanding of the impact of flooding when flood protection measures are absent or fail.

Streamlining Information Utilizing Knowledge of Perceptions

Knowledge about perceptions of the hazard by different groups is indispensable for assuring that information on floods really reaches the target and is understood. The perception studies within the FLOWS Project offer an improved basis for dissemination of such information indicating what kind of information is required and also the most effective links for spreading it.

Poor awareness of laymen about the flood hazard calls for better information about this natural hazard that cannot be eliminated. Information on how flood loss can be reduced (e.g. retrofitting, improved spatial planning, better use of warning systems etc.) might at the same time counteract "the will of God" attitude helping people learning how to live with floods. Such information proved to be specially appreciated and considered as more relevant if it came from local authorities. Alternative ways of disseminating information were also demonstrated within the project (campaigns, mobile information centres etc.).

Citizens involved in the poll study demonstrated a rather poor knowledge about flood management and defence measures in their communities. One of the suggested ways to assess this is using school children as a bridge for spreading information on flooding and flood protection between the authorities and laymen. The advantages with using this information channel were stressed during national and international expert panels. The school projects undertaken in the FLOWS Project have gotten a lot of attention in the local community and in the press stimulating interest in the flood topic.

A severe underestimation of the flood hazard revealed by the poll investigation indicates that the information available (like "100-year flood") is misunderstood and new ways of presenting such information should be looked for, ways that are easier for an individual to have associations with. As noted by Renn (2004), it is not only probability and severity of adverse effects that influence the way people perceive risk but rather the context in which the risk is experienced.

A rather widely spread mistrust in the way public authorities handle flood issues suggests a necessity of an alternative information link viz. from people with "flood experiences" to those facing a flood hazard (e.g. newly moved to a flood prone area) or flooding for the first time. "The National Flood Forum", UK has already accumulated much experience of such information transfer (Dhonau, 2005). This communication link also promotes active citizenship.

Findings from the focus groups indicate that people might find it easier to engage on behalf of others (i.e. children and elderly). Tuning information towards the security of vulnerable groups of the society might stimulate willingness to learn more about flooding as well as participate in flood assessment activities of the community.

Also decision makers expressed a clear wish for better and easily accessible information on flooding. Especially appreciated were flood hazard maps. When available, such maps produced by different authorities sometimes appear to be incoherent, which significantly reduces their value for decision making. Coherence and coordination of all information on floods propagated to decision makers and population should be better prepared in future.

Among the decision makers a possibility of direct contacts with their colleagues nationally and internationally was much appreciated. Experts came with many suggestions for improving flood safety. Sharing great experience accumulated in different countries in combat with flooding offers a valuable information source that is underestimated at present. The expert panels, like those organised in the frame of the FLOWS Project, deserve to become a natural component in flood assessment.

Conclusions: Engaging a Dialogue between Laymen and Decision-Makers

The ambition within the part of the FLOWS Project that focus on perception of the flood hazard is, besides bringing valuable new data on perception in the North Sea region of Europe, to show how this information can be accommodated in flood assessment. In this paper we showed how such information may be used for streamlining information on flooding. The poll study of flood hazard perception was a first step on the way. The focus groups studies that followed the poll allowed a deeper investigation of a number of selected topics covered by the poll. The views of decision makers on flood hazard were studied by means of expert panel discussions at national and international levels.

Both poll and focus groups studies revealed an underestimation of flood risk and passive attitudes towards flood problems. Only less than half of respondents really knew that they lived in an area at risk of flooding, which can partly be attributed to inadequate information. False feeling of safety given by ignorance about the flood hazard or over-belief in the protection provided by technical defence measures demonstrated by laymen witness that information on floods was misunderstood or neglected. Better information formulated in a way that is easily understood and interest general public is obviously required. Within the project a number of different disseminating techniques have been demonstrated: school projects, awareness campaigns, information desks, leaflets to inhabitants, improved flood warning and new ways of visualising flood risk.

The alarming fact that many citizens lack confidence in the ability of public authorities to handle flood problems demonstrates the existing gap between the decision makers and general public. Though this can partly be explained by passive attitudes among laymen, the majority of decision makers generally do not yet attribute crucial importance to public participation in flood assessment and neither to the regular information on flooding. In this respect the positive attitudes expressed by the majority of the British experts might serve an example for the others.

Passive attitudes towards taking actions to lower the adverse effects of flooding on the individual level may be effectively assessed by engaging people with flood experiences for spreading information about the flood hazard and about what can be done to adapt homes. This way of information dissemination becomes still more effective when the advice is given by citizen organisations, which can at the same time represent laymen interests in contacts with the authorities. A positive example of "The National Flood Forum" in the UK encourages promoting this experience also in other countries.

In this study we let two different perceptions, i.e. decision makers and laymen meet, and then engage a dialogue between these two parties trying to reach a consensus on what risk is tolerable and what protection is possible to provide. The experts from different countries and representatives from two public organizations that are engaged in water problems met at the international panel to exchange views on what risk can be tolerable, suggest approaches to trigger active public participation, and also suitable communication links between decision makers and laymen living in areas at risk. The dialogue decision makers—laymen have been launched as an important element towards governance, which includes decision making and tradeoffs, stakeholder participation, policy etc. This dialogue may then continue with a "trialogue" – government, private sector and civil society (Falkenmark, 2004). As correctly noted by Renn (2004), "public perception and common sense cannot replace science and policy but they can certainly provide impetus for the decision-making process".

References

Environment Agency, 2004. Flood Warning Dissemination. National Awareness Survey.

Falkenmark, M., 2004. Symposium Conclusions, 2004 Stockholm Water Symposium. *Stockholm WaterFront,* **3,** September 4-5, 2004.

Dhonau, M., 2005. The National Flood Forum. FLOWS Third International Conference, January 11-12, 2005, Conference Report—external.

Enander, Ann, 2005. Risk perception in a wider context: how do the results from FLOWS fit with other studies on risk perception? FLOWS Third International Conference, January 11-12, 2005, Conference Report—external.

EM-DAT, 2004. The OFDA/CRED International Disaster Database, www.em-dat.net. Univ. Catholique de Louvain, Brussels, Belgium.

FLOWS, 2005. FLOWS Third International Conference, January 11-12, 2005, Conference Report—external.

Horven Skellnes, J.L., 2001. "Med fare for flom..." (In danger of flooding..., in Norwegian). Georgafisk Institutt, NTNU, Hovedfagsoppgave.

Krasovskaia, Irina, 2005a. Perception of Flood Hazard in Countries of the North Sea Region of Europe (Draft Report): Results from the Poll Study.

Krasovskaia, Irina, 2005b. Perception of Flood Hazard in Countries of the North Sea Region of Europe (Draft Report): Results from the Expert Panels.

Krasovskaia, I., Gottschalk, L., Sælthun N.R. and Berg, H., 2001. Perception of the risk of flooding: The case of the 1995 flood in Norway. *Hydrol. Sci. J.,* **46** (Special Issue): 855-868.

Kundzewicz, Z.W., 1999. Flood protection—sustainability issues. *Hydrol. Sci. J.,* **44(4):** 559-571.

Morris-Oswald, M. and Simonovic, S.P., 1997. Assessment of the Social Impact of Flooding for Use in Flood Management in the Red River Basin. Report prepared for the International Joint Commission Red River Basin Task Force, Winnipeg, Canada.

TNS Gallup, 2004. *Fokusgrupper i Norge, Notodden og Åsnes kommuner* (in Norwegian).

Renn, O., 2004. Perception of risks. *Toxicology Letters,* **149:** 405-413.

Rosslyn Research, 2004. FLOWS Focus Groups -Flood Risk Perceptions, Inland East Anglia, Conducted for Cambridgeshire County Council, Job Number 3262.

Tønsberg, Sigurd, 2005. Active citizenship in Planning Process. FLOWS Third International Conference, January 11-12, 2005, presentation.

World Disaster Report, 2004. Red Cross Red Crescent.

20

Decreasing the Risk of Floods in Small and Medium Sized Catchments through Natural Storage in Headwater and Riparian Zones

Roelof J. Stuurman, Perry G.B. de Louw and Marc F.P. Bierkens
TNO Built Environment and Geosciences, Utrecht, The Netherlands

Introduction

In many small to medium sized catchments (up to 500-1000 km^2) in Europe, natural storage in both headwaters and riparian areas has been steadily decreasing by large scale drainage and urbanisation. An example of decreasing storage in head water zones is given in Figures 1 and 2, where for a small catchment in the south of the Netherlands the original areas of lakes and heathland on the catchment divide are shown. Also shown is the decrease in area of moorland pools since 1850.

In these natural river systems much of the rain water was captured by the depressions (forming moorland pools) and heathlands. This water subsequently evaporated or infiltrated to contribute to groundwater recharge. In these systems a considerable part of the runoff thus originated from baseflow and large rainfall events were levelled off considerably. During high discharge events additional storage was available in the riparian zone through inundation of the floodplain.

From the end of the 19th century the moorland pools were drained by connecting them by a network of ditches (Figures 1 and 2). At the same time the moorlands and heathlands were drained and turned into agricultural land. Also, in the riparian zone the rivers were constrained to keep them from meandering and embanked to decrease inundation frequency. These activities resulted in a river system with a considerable faster response to precipitation. If climate is changing, causing more precipitation as well as a higher probability

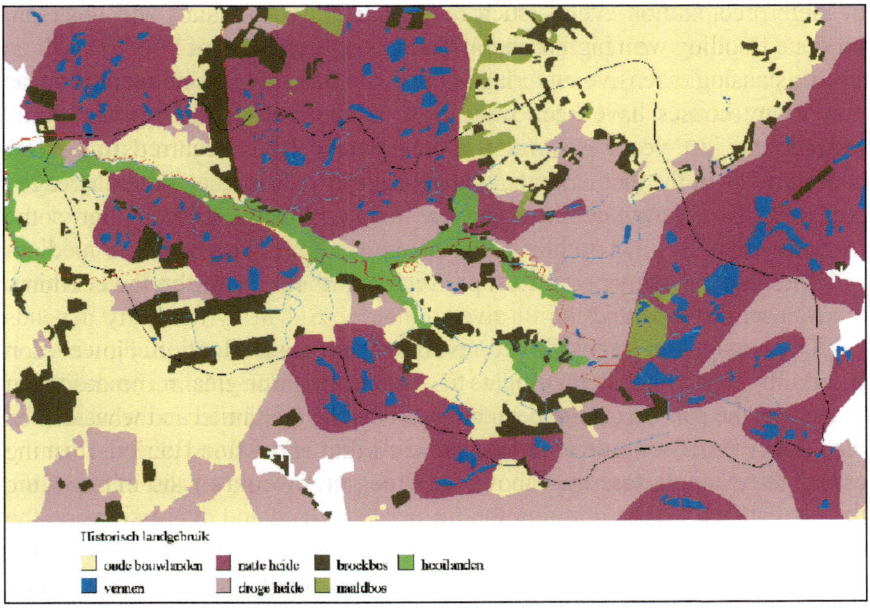

Figure 1: Historical map showing a natural brook and river system in the 19th century. A considerable part of precipitation water is captured in depressions (forming lakes) and heathland; green: river valley (riperian area), purple: heather, blue: lakes, yellow: agricultural land.

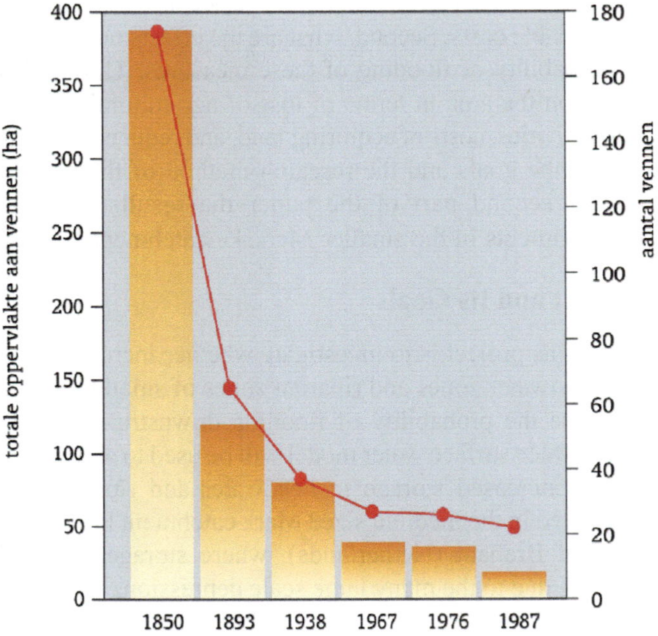

Figure 2: The area of lakes in the small catchment of "het Merkske" (about 4000 ha) has been decreasing considerably since 1850.

of large precipitation events, such fast reacting river systems will therefore produce flooding with high probability. The consequences of flooding will be severe, causing extensive material damage and possible loss of human life, as now many houses have been built close to the river in the riparian zone. Indeed, the last years' several flooding events have occurred in several communities in the Netherlands, such as the city of Breda at the outlet of the medium sized Mark catchment and the town of Meppel at the outlet of the Meppelerdiep. Similar problems can be expected in many small to medium sized lowland catchments in Europe, such as tributaries to the lower Rhine, the Meuse, the Elbe and the Po river.

Obviously, it is impossible to revert to the natural situation. However, it is certainly worthwhile to investigate to what extent the original accommodation space (i.e. the part that is not been built on) can be reclaimed and what will be the effect of such measures on mitigating current and future flood risk. In the case of the example described above, such measures would consist of removing the network of ditches draining the former heathland and depressions on the divide, as well as allowing part of the river in the riparian area to inundate and meander again. In other catchments (e.g. the Meuse) increasing storage would be focused on the headwaters themselves, for instance by removing embankments in the upper reaches of the river in order to stimulate meandering and multichannel flow. The hope is that, if such measures are taken at a large enough scale, the probability of flooding may be significantly decreased.

Several questions should be answered before one would attempt such measures. First, how much accommodation space is still left and can be reclaimed at reasonable costs. Second, what are the effects on the hydrological response and probability of flooding of these measures. Third, what are the economic effects on the area in terms of loss of agricultural production and other economic activities, costs of acquiring land, and reduced risk of flooding.

In this article the goals and the research method of the project will be explained. In the second part of the paper the results of groundwater conservation experiments in the smaller Merkske catchment are presented.

The Mark Project and Its Goals

The main goal of this project is to investigate whether increasing the storage of water in the headwater zones and riparian zones of small to medium sized rivers will decrease the probability of flooding downstream. To this end a combined groundwater-surface water model will be used to simulate discharge with and without increased storage in headwater and riparian zones. The model will be applied in the medium sized Mark catchment in the western part of the province of Brabant (Netherlands), where storage possibilities are expected to be high due to the many large scale depressions on its divide. The transboundary Mark catchment has an area of 76,650 ha. The upper part of the catchment (50%) is situated in Belgium. Apart from a hydrological analysis, the feasibility of storage measures will be analysed, as well as economic costs and benefits, the consequences for the spatial quality of the area and local support for such measures. Also, it will be investigated whether increasing headwater storage, and thereby baseflow, will have a positive effect on the groundwater quality in the downstream river valley.

Research Plan

The following steps are taken to reach the project goals:

1. A historical-geographical survey will be conducted to assess the occurrence of depressions and heathland in the headwater zone and the areas of inundation in the riparian zone. The associated accommodation space for water will be estimated.
2. A combined groundwater-surface model will be built and calibrated to the present situation.
3. Using historical time series of climate variables and discharges as well as the results from the historical-geographical survey, the model is used to reconstruct the discharge regime and groundwater depths around 1850.
4. The hydrological model is used to determine the hydro(-ecological) effects on the groundwater system and the surface water system after implementation of planned reconstruction projects in the (transboundary) catchment.
5. The hydrological model is used to seek for possibilities of increasing water accommodation space under the current situation. This can be done by running scenarios with different levels of adaptation of land use and drainage density. Each scenario will lead to a different hydrological regime with probabilities of flooding as well as different groundwater levels. Results for each scenario will be derived under the present climate, as well as under expected climatic input in 2050 and 2100. Extreme scenarios are the reconstructed situation around 1850 and doing nothing. For each scenario the geographic and economic consequences will be evaluated. As a side product, each scenario will also lead to an estimate of groundwater quality in the riparian zone, which is an important indicator for nature development in this area.
6. Each scenario will be presented to stakeholders (landowners, farmers, property owners, provinces, water boards and municipalities), such that through discussion and a participatory process, suitable alternatives can be identified. Selected alternatives are subsequently presented to a broader audience to gauge public support for the proposed measures of flood mitigation.
7. Field experiments will provide information about the hydrochemical and ecological effects of restored flooding in the moorland pools and brook valleys.

The study focuses on the entire (76,650 ha) Mark catchment. Also the detailed integrated models for unsaturated zone, groundwater and surface water will cover this catchment. Most part of the experimental field studies will be carried out in the smaller Merkske catchment (6000 ha). The Merkske is part of the Mark catchment and is also equally divided between Belgium and The Netherlands. In this case, the effects on the groundwater system of measures in the surface water system were calculated and measured. When more groundwater is conserved, less discharge of surface water will happen.

Results: Merkske Catchment Experiments

Groundwater conservation leads to higher phreatic groundwater levels and less peak discharge of surface water. Conserving groundwater in the headwater region is the first step in taking measures to reducing the risks of flooding. When all attempts to conserve the precipitation surplus as groundwater are taken, it's time to store the water in depressions such as moorland pools or in the riparian zone.

In the Merkske catchment the focus was to conserve groundwater by reducing the drainage capacity with measures such as raising ditch bottoms, filling in ditches or installing weirs. Several field experiments were executed in order to measure the effect on the groundwater level. Also a groundwater model was used to calculate the effects of the different surface water measures.

Weirs as a Water Conservation Measure

The positioning of weirs in the higher agricultural area is based on the philosophy that an attempt should be made to retain the water in the area at the highest possible level. In this way the water shortage for the crops can be reduced so that less irrigation is required. By increasing the phreatic groundwater level in the infiltration area and decreasing the groundwater abstraction for irrigation, the groundwater head in the deeper aquifer will also increase. This will enhance the deep groundwater discharge in the brook valleys, which will benefit nature.

Dehydration of these natural areas (decreasing deep groundwater discharge and decreasing groundwater level) has led to acidification of the root zone (Figure 3) and mineralisation of the peat which in its turn has led to extra nutrients being deposited. These are the most important causes for the decreasing natural values in the brook valley. It was hoped that the positioning of the weirs would stop the deterioration of these wetlands.

However, measurements showed that the effect of the weirs was limited. No influence on the phreatic groundwater level could be measured at a distance of 50-100 metres from the watercourse. The models showed that the effect of weirs on the phreatic groundwater level is very small and is quite restricted. After 100 metres it is no more than 5 cm at a surface water level increase of 70 cm (see Figure 3), at an increase of 30 cm the effect has already completely

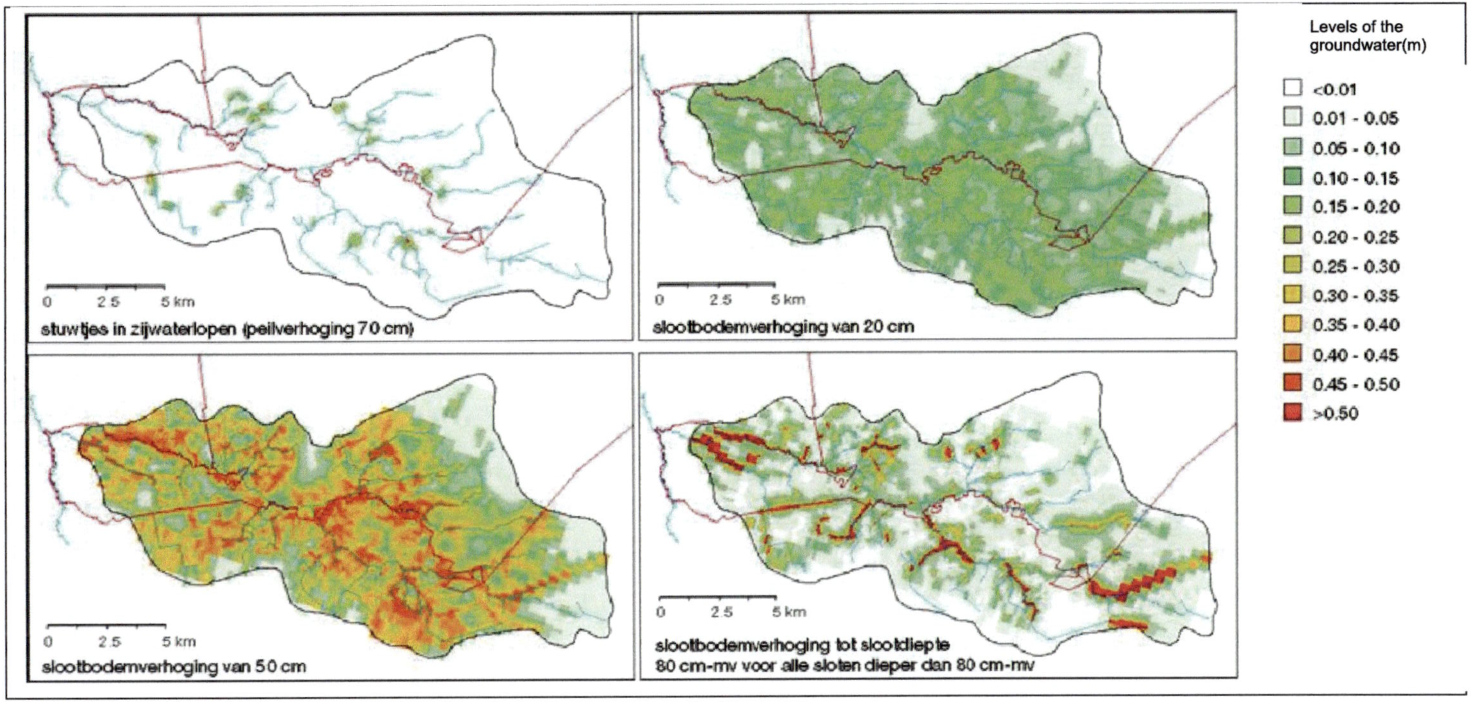

Figure 3: The effect on the phreatic groundwater level of various surface water measures, calculated with the regional time dependent groundwater model (MODFLOW). The classification presents the increase of groundwater level (m), above (left) presents the effects of weirs (0.7 m), above (right) the effect of increased level of the bottom of ditches (0.2 m), below (left) diem (0.5 m), below (right) increase of all surface water levels to 0.8 m below surface level.

disappeared at a distance of 50-100 metres from the watercourse. The groundwater head in the deeper aquifer is not influenced by this small rise in the phreatic groundwater level in the higher agricultural areas and consequently the deep groundwater discharge in the brook valleys does not increase either.

Also a time effect was observed in the model results. The maximum effect of the weirs was not achieved until two years after their installation and it appeared to be much smaller in summer than in winter. Drainage through the watercourses hardly plays a role in summer so only a small amount of water can be retained by a surface water level increase. Moreover, the weir experiments showed that sometimes the desired surface water level could not be obtained because of limited discharge through the watercourse.

The limited effect of the weirs can be explained by the relatively large gradient of the watercourses in the area. A 70 cm surface water level increase has completely disappeared at a distance of 400-500 metres upstream. Moreover, the phreatic aquifer with its many watercourses and ditches is thin and the permeability is small, so that effects are only noticed at short distances.

Measurements in the brook valley showed that the effect of a weir strongly depends on the type of soil through which the brook flows. In the Halsche Beemd the Merkske brook flows through peat. As a result of a brook level change of 75 cm in April 1997, a small effect was measured on the phreatic groundwater level of 40 cm at a distance of one metre from the brook, 23 cm at a distance of three metres, 13 cm at a distance of 10 metres. At a distance of 20 metres from the brook no effects were measured (see Figure 4).

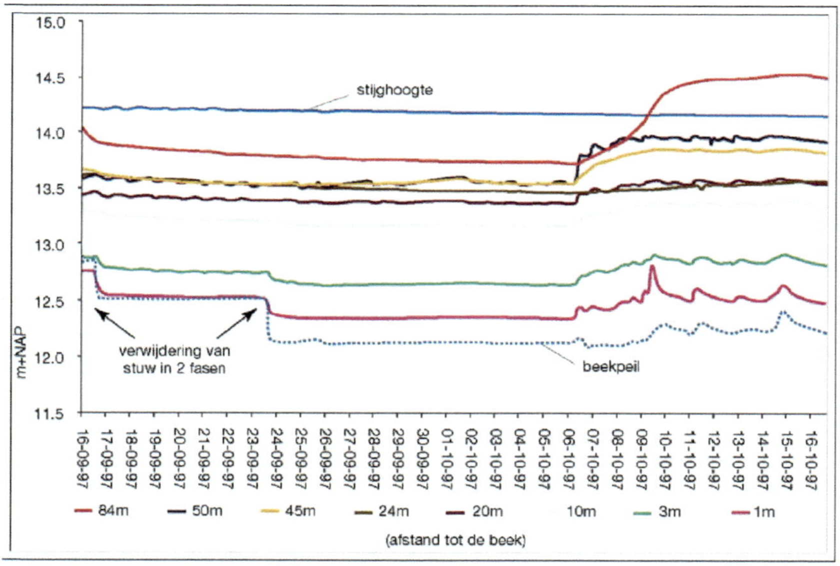

Figure 4: The effect on the phreatic groundwater level in the Halsche Beemd of a sudden drop of the stream level (0.75 m), measured with 'divers' [stijghoogte = hydraulic head; beekpeil = surface water level; verwijdering van stuwen in 2 fasen = removel weirs at two time steps; afstand tot de beek = groundwater level at a certain distance from surface water]

Raising Ditch Bottoms and Filling in Ditches

In addition to the effect of weirs, the effects of raising ditch bottoms and filling in ditches on the phreatic groundwater level and the groundwater head were also considered. These effects were also calculated with the regional groundwater model (Figure 5). It appeared that raising the bottoms of all ditches by 20 cm leads to a phreatic groundwater level increase of 5-20 cm throughout the entire catchment area and a groundwater head increase of 8 cm. Raising the ditch bottoms by 50 cm, has a proportionally larger effect. Filling in all watercourses and ditches leads to a groundwater head increase of on average 20-25 cm. This situation corresponds to the historical situation previous to the cultivation of the heathland. So since that time the groundwater head has decreased by 20-25 cm, as a result of digging watercourses and ditches for the cultivation of the area.

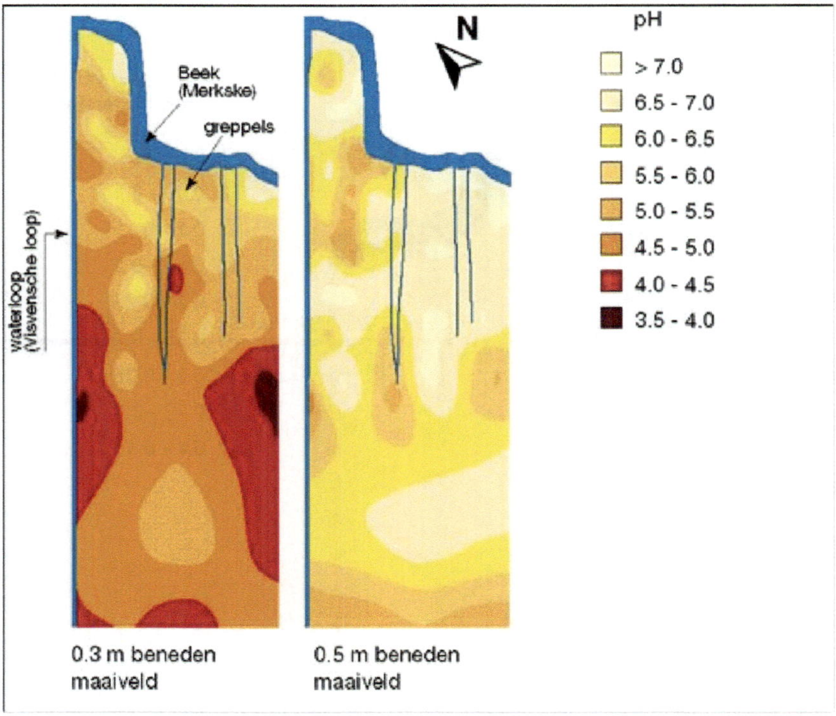

Figure 5: Bird's eye view of the Halsche Beemd hayfield in the brook valley of the Merkske stream. Spatial distribution of the soil pH at 30 cm and 50 cm below surface level. The upper 30 cm of the soil is acidified due to dehydration. (beek = brook, greppel = small ditch, waterloop = ditch, 0.3 m beneden maaiveld = 30 cm below surface level). The area is 100 × 40 metre.

Effects in the Riparian Zone

With the help of the regional groundwater model it has been calculated that the groundwater head in the deeper aquifer underneath the Halsche Beemd has dropped by about 0.85-1.00 metres since the cultivation of the (wet) heathland and the beginning of the groundwater abstractions. 60-70% of this is due to groundwater abstractions and 30-40% of it is caused by the system of watercourses, ditches and trenches created since then. Recovery of the natural/historical groundwater head seems unrealistic, although it would be very favourable for the ecological values in the brook valley wetland system. With the local Halsche Beemd model it has been calculated that an increase of the groundwater head by 100 cm would create an increase in deep groundwater discharge as well as an increase of the phreatic groundwater level of an average of 20-30 cm over the entire area (Figure 6). The low-nutrient, iron-

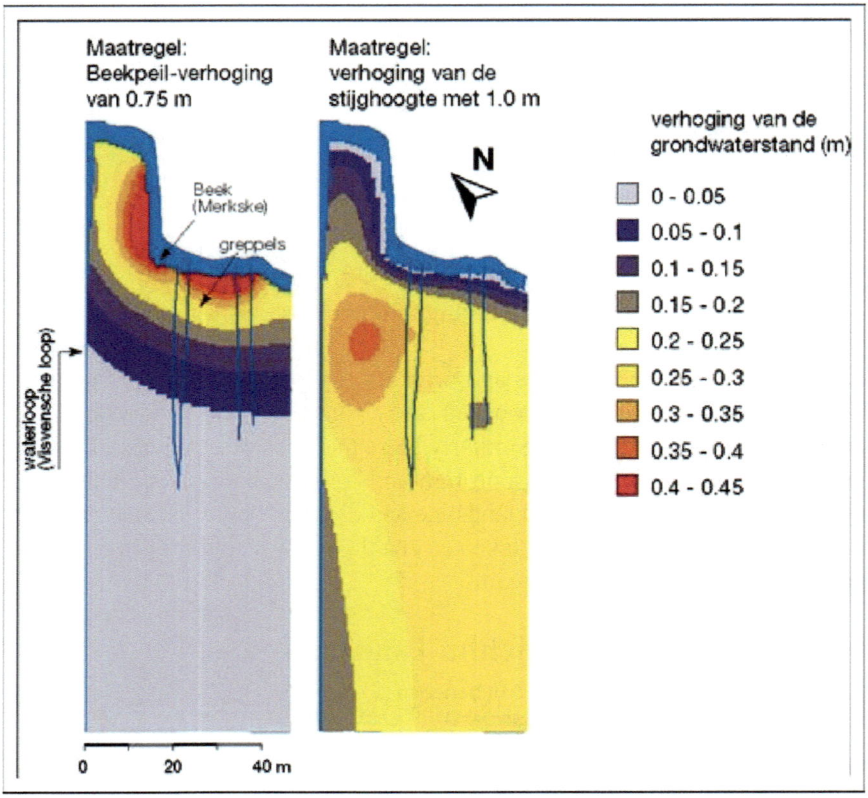

Figure 6: Bird's eye view of the Halsche Beemd hayfield in the brook valley of the Merkske stream.The effect of respective stream level and groundwater head increase on the phreatic groundwater level, calculated with the local groundwater model. [verhoging grondwaterstand = increase in groundwater level (m); maatregel:beekpeilverhoging van 0.75 cm = increased surface water level of 75 cm; maatregel: verhoging van de stijghoogte met 1.0 m = results of increased hydraulic head of 100 cm)

and calcium-rich deep groundwater discharge would then reach up to the root zone of the groundwater dependent vegetation again and the influence of the acid rainwater diminishes. Because of this the upper 20-30 cm of the soil profile (root zone) will become more alkaline again. Increasing the brook level by 0.75 metres only affects the phreatic groundwater level in a zone of 30 metres along the brook (Figure 6), which was confirmed by the results of the weir experiments (Figure 4).

Both historical and soil research showed that in a natural situation the brook valley used to be flooded for long periods of time during the winter. This flooding stopped due to brook improvements in the years 1960-1970.

Conclusions of the Merkske Experiments

Groundwater-oriented surface water level management with the help of weirs only offers minimal contributions to water conservation in this sandy area. The most important reasons for this are: (1) the hilly terrain, (2) the relatively poor permeability of the cover sand, which causes the groundwater to react slowly to a surface water level change, (3) the fact that the weirs may only be installed when the 'winter water' has largely been discharged and (4) the main watercourses are situated in naturally low areas where no water shortage occurs in summer. Weirs in the watercourses have too small an effect on the phreatic groundwater level and no effect at all on the groundwater head, so that nature in the brook valley does not benefit.

In the brook valley itself groundwater-oriented level management may be useful at some places. In that case the most important criteria are: (1) the surface of the flat area, (2) the shallow geology and (3) the functional use.

Optimisation of the drainage system of ditches is a good alternative. Many of these are over-dimensioned and have lost their original function or have become redundant. Over the past hundreds of years the drainage system has evolved into a totally inefficient system because of individual, location-specific actions. It was proposed to optimise this system by considering in the field, starting from the infiltration area, which ditches can be filled in or where ditch bottoms can be raised. In this way the infiltration area is enlarged, the winter water is retained higher up in the area and the groundwater head will increase, which in its turn will cause the deep groundwater discharge in the brook valley to increase.

The study also showed that the groundwater situation in the brook valley is much more dependent on the groundwater head/deep groundwater discharge than on the brook level. Recovery of the (regional) groundwater head and the deep groundwater discharge is essential to recover the ecological value of the wetland system. The groundwater head has decreased 85-100 cm as compared to the natural situation. Groundwater abstractions for drinking water and irrigation account for about 60% of this decrease, drainage for about 40%.

When the intense drainage system is extended by filling in ditches of raising ditch bottoms, more rainwater is conserved as groundwater. The discharge of this conserved groundwater will be more evenly distributed in time. In periods of heavy rainfall less discharge will take place and the risks of flooding downstream will be reduced. In dryer periods the baseflow will increase. When all the attempts of conserving groundwater are done, then the excess of water must be stored in depressions or in the riparian zone. How and where storage of surface water will have the best effects, will be investigated in the Mark-project.

Reference

De Louw, P.G.B. and Stuurman, R.J., 1999. Regional and local hydrological influences on a brook valley wetland system. *In:* Means, J.L. and Hinchee, R.E. (Eds.) Wetlands and Remediation. Batelle Press, Salt Lake City.

21

Estimating Sediment Mobilisation from Torrent and Gully Deposits: Field Studies

Martin J. Haigh
Department of Anthropology and Geography
Oxford Brookes University, Oxford, UK

Introduction

The bed-load sediments that are found trapped in the depositional alluvial fans of ephemeral gullies and ravines are the remains of sediments mobilised by erosion in the catchment minus those sediments removed by flowing water. Erosion is selective; it removes different parts of the soil and regolith in different proportions and the least easily moved sediments dominate the sediments trapped in depositional fans. Although the rules which govern selective erosion are quite complicated, there remains the possibility of measuring the volume and particle size distribution of the sediments in an alluvial fan and use this information to calculate the total amount of sediment removed by the flows that created the fan.

In fact, this approach has been employed by generations of landslide researchers and by those who study debris flows, which can be 5-76% coarse sediment (>64 mm) (Haigh et al., 2004; Webb et al., 2000). This study explores some of the problems that emerge when an attempt is made to extend this logic to the analysis of deposits created by flows in first order torrents and gullies. Of course, if it were possible, easily, to perform this calculation, then it would be equally easy to determine how much sediment a particular ephemeral basin contributed to the general sediment load of an affected river basin during a particular extreme event. Unfortunately, the problem is not so simple (Walling, 1988). This article reviews some of the issues.

Particle Size Analysis

Particle size analysis, like Munsell Soil Colour scores and pH, is a soil property more frequently measured than put to good use. True, it has major applications in discussions of soil aggregate stability and also in soil loss prediction, where it is a factor in the USLE, WEPP and many other technologies. This article deploys particle size analysis in a study of the processes of selective erosion and considers the extent to which particle size data of mobilised sediment can be used to estimate sediment yield. It is founded in the simple laboratory techniques of wet sieve and hydrometer analysis (Loveland and Whalley, 1991). It reinvents an intuitively obvious but little heralded approach to soil loss estimation, here graced with the title: 'The Lost Soil Method'. This is evaluated as a tool in studies of some intensively monitored micro-catchments in the USA and India.

Himalaya Case Study

First, India and a study based on measurements of the accumulation of alluvial sediments from four, parallel, adjacent, first-order, torrents on a steep, south-facing, ridge in the Lesser Himalaya. Two of these streams flow through areas that retain their forest cover and two through areas where the major trees have been removed and replaced by a goat-ruptured turf and thin scrub.

The site is at Landour, near Mussoorie, Uttaranchal (30.27 N, 78.60 E). The geology is Tal Formation shale and sandstones, which dip southwards (160-210° T.B.) at angles of 20-30° (Range: 8-52°). The hill-slope is very steep, ground-surveyed in 2 m unit lengths, 92% is steeper than 15° and 41% steeper than 35°. Forest covers 55%, agricultural terraces just 2% and housing and roads about 34% of the land area (Haigh, 1979). All four catchments include houses and are crossed by the two paved roads that run east-west along the contour. Annual rainfall is 2356 mm—but this study was undertaken after a twice-normal intensity monsoon and nearly all the rain fell in the period July-September. Frost action is slight. Winter snowfalls are short-lived.

The four micro-catchments explored for this study have similar physical characteristics. They have a local relief of approximately 300 m, their areas range between 0.8 and 1.1 km^2. Their single main channel is rock floored and slopes at about 40°. Channel walls of soil and weathered rock slope at 50-70° and are major sources of sediment. However, the two channels that drain the deforested slopes are much wider, deeper and show greater bed-load than those in the forest.

A bed-load sediment trap is an obstacle introduced into a stream channel's long profile that causes a large portion of its sediment load to be retained. In this case, such a trap was created accidentally by road works on the lower of the roads that cross the catchment. Culverts were blocked, so in monsoon flood conditions, streams flowed right across the road-bed where, the abrupt change of angle caused them to deposit sediments and build up alluvial fans, which, for these four streams, were entirely contained on the road-bed.

In a hiatus between the monsoon and road clearing works, the size of these alluvial fans and the calibre of the season's alluvial fan deposits were measured. Samples collected were subjected to particle-size fractionation by technical staff at the School of Environmental Sciences, Jawaharlal Nehru University, New Delhi.

Despite their similarities in slope and area, the forested and deforested catchments generated very different sized alluvial fans. The deposits created by the streams flowing from the forest amounted to about 48 m^3, the equivalent of a layer 0.5 mm deep across the catchment. The deposits beneath the deforested lands were much larger: 230 m^3 and 370 m^3 of debris, the equivalent of surface losses of 2.8 mm and 3.7 mm.

These volumes, of course, do not represent total sediment transport. It is certain that much debris was carried beyond the road-bed alluvial fan in suspension and solution. In areas suffering recent deforestation and development, small channels have transformed from small perennially flowing rock floored streams to dry boulder choked swaths where surface flows only occur in monsoon spate, all within living memory. However, all that can be said here is that the volume of sediment trapped beneath the forested slopes is 5 to 7 times smaller than that trapped beneath the slopes which retain the bulk of their forest cover. Meijerink (1974) estimated that the natural rate of ground loss in the neighbouring, mainly deforested, Aglar catchment as 5.9 mm/year, so these results imply that natural rates of ground loss should be closer to 1 mm/year.

However, placing the particle size analysis curves one over the other for the four catchments and setting them alongside the curves for the channel wall sediments makes it quite obvious which sediment has gone. The finer fractions are least well represented and, of course, the most coarse fractions are least changed. This raises the tantalising possibility of matching together the proportions of the soil and the alluvial deposits in the larger sediment classes, use this to calculate the proportions of each smaller size class that have been removed, and extrapolate these totals to calculate how much soil and subsoil would have been needed to generate the amount of sediment deposited. This would provide a way of calculating total sediment discharge from the calibre of the debris in each alluvial fan.

This sounds like a very good idea. However, there are many problems. First, it is difficult to sample the larger size fractions in both soil/subsoil and alluvial sediment. Sample size is a function of the size of the calibre of sediment of interest and that would require a major screening operation. Estimations of the amount of sediment needed to sample for a maximum particle size of 180 mm range upwards from 80 kg (Bunte and Abt, 2001; Lenzi and Marchi, 2000). It would not be possible to collect and transport such large samples from the field, given the resources available.

However, the results of the particle-size analyses provide some evidence of the way in which sediment transport differs between the two types of catchment. Particle-size fractionation was undertaken for samples of the sediments accumulated in each of the alluvial fans and for soil samples derived from the eroding areas of each catchment. It was found that the soils in both the forested and deforested catchments contained a similar proportion of silt and clay. The diameter of the smallest 10% of debris (D_{10}) in all of the soil samples was 0.2 mm or less. However, the soils in the forested catchments contained a far higher percentage of sand-sized particles. The mean particle size for the forest soils was 2 mm whilst that for the soils in the deforested catchments was 4 mm and above. By contrast, the debris in the small accumulations beneath the forested catchments was very coarse (D_{10}: 5.0 mm and above), whilst that in the larger accumulations beneath the deforested catchments contained a relatively high proportion of finer material (D_{10}: 0.7 mm, D_{50}: 5.0 mm). These results indicate that part of the increased sediment production in the deforested areas was due to an increased mobilization of sand and fine gravel-sized particles.

Selective Erosion

Observations of eroding soils, of course, always demonstrate selective erosion. The particles that are removed and which discolour field run-off are fine particles produced from the disruption of unstable soil surface aggregates, often dominated by silt-fine sand and clay sized particles. Larger particles are left behind either at the soil surface or in the bed-load of rills and gully channels.

Unfortunately, particle size is not the only factor at work. Table 1 compares data from some abandoned surface mine spoil dumps at Henryetta, Oklahoma, USA (35.44N 95.98W). These were created by the activities of two strip-mine operations: an Alkonak mine between 1916 and 1924 and a McGinnis and Grafe mine from 1948 until the late 1950s. The mines employed the same technique to strip the same coal seam buried beneath the same amount (10-12 m) of similar overburden (D_{50}: 0.1-1.0 mm). The two sites are contiguous, have similar surface topographies, but one has been subjected to 40 years more weathering and erosion than the other. Two typical unvegetated, east-facing, slopes were examined. They both had a local relief of 3.5 m, a length of 8 m, and a three-metre-maximum slope angle of approximately 34°. One slope was situated on the older mine dumps, the other on the younger mine dumps (Goodman and Haigh, 1982). Replicated samples were taken from the top 200 mm of soil at five sites on each slope and also from the less-weathered layer at 0.6 m depth. Particle size analysis of these deep samples found larger mean particle sizes on the younger than older slope, indicating active breakdown, but D_{10} for both was in the clay size range. The surface samples were then washed through a 6 mm sieve and the amount retained sorted by rock type and weighed. Finally, a sample of 50 typical fragments of each type was measured to determine their flatness and bulk density (Bunte and Abt, 2001).

Table 1: Characteristics of soil skeleton (Fragments: >5 mm)

		Weight (g/kg)		
Stone Type	*Bulk density (g/cm^3)*	*Younger Slope (sd)*	*Older Slope (sd)*	*Significant difference (p=)*
Coals	1.3	507 (32)	156 (51)	0.0001
Shales	2.4	234 (67)	289 (41)	ns
Sandstones	2.6	227 (37)	511 (47)	0.0001
Ironstones	2.6	32 (48)	44 (57)	ns
		Flatness Index ((a +b) /2c)		
Coals	1.3	3.6 (0.7)	3.6 (1.8)	ns
Shales	2.4	5.6 (1.9)	4.4 (1.2)	0.01
Sandstones	2.6	3.4 (0.5)	2.9 (0.8)	ns
Ironstones	2.6	5.5 (0.7)	4.8 (1.0)	0.03

The results (Table 1) show selective erosion at work. Low density coals are selectively reduced in the surface samples on the older slope profile. The proportion of poorly consolidated shale and mudstones, which break down very rapidly when subjected to weathering, changes little. By contrast, the proportion of durable sandstone fragments increases significantly.

The lower part of the table shows the role of particle morphology in erosion using a Flatness Index, which is simply, the sum of the two major axes divided by the minor axis for each fragment. The argument is that more platy fragments will be more easily rafted off by runoff. There are fewer platy fragments in the surface spoils of the older site, especially for the more platy shale and ironstone fragments.

Road Cut Micro-Catchment

To the west, lie the Central Redbed Plains of Oklahoma. There are >7.0 million hectares of road-banks in the USA, many suffer from erosion, many contribute significantly to the sediment pollution of adjacent lands. This study considers a micro-catchment within a typical, small, road-bank that flanks a two-lane, rural highway in Cleveland County (35.22N 97.44W (Sec 19, T.9.N. R.1.E)). This microcatchment is three metres high, has a ground surface length of 8.5 metres and a mean slope of 27° (51% grade). It is typical of such road margin features throughout the world's savannahs, but this site is selected because, for five years, it was the subject of intensive measurement associated with study of slope evolution and microrills (Haigh, 1984).

The road-bank was created around 1965 and is cut into a gently sloping (3-5°) upland on the edge of semi-natural 'post oak' woodland. The sparse canopy of this scrubby woodland allows a dense ground vegetation dominated by Bluestem grasses (50% composition dry weight). Forbs account for just 10% of under-storey biomass (Bourlier, 1976). However, the road bank itself is mainly unvegetated with no more than an occasional grass tuft colonising the gully channel floor.

The subtropical climate has a rainfall of 875 mm/year. The area is called 'Tornado Alley' and the heaviest rains are caused by convectional storms in the later spring months, which ensure high rainfall erosivity, the USLE's R-factor is about 250. Temperatures are high, the July average is about 27 °C, January about 3 °C. There may be 65 winter freeze-thaw cycles each year although the freezing wave penetrates no deeper than 100 mm.

The road cut is capped by a loamy fine sand (Darsil Series) soil, which is a beheaded Udic Haplostalf of about 480 mm depth. It is highly permeable (50-500 mm/hr). Its Plasticity Index is about 7, its Liquid Limit less than 26, its shrink-swell potential is low, and its soil erodibility (K) factor from the Universal Soil Loss Equation (USLE) is about 0.2 (Bourlier, 1976). It outcrops across the upper two metres of the 8.7 metres road-bank surface and channel incision begins with the base of the soil. Annual average ground loss from the non-gully slopes shows a strong correlation with total annual precipitation ($r>0.9$) and with precipitation received in daily totals of more than 12.5 mm ($r>0.96$).

Garber Sandstone bedrock outcrops beneath the subsoil over a distance of 0.1 metres on the interfluve and 0.4 metres in the floor of the gully channel. This is poorly cemented, fairly coarse grained, red sandstone (Wood and Burton, 1968).

Down slope of the bedrock and underlying >4 metres of the road-bank surface from the outcrop to the upper limit of the 2-3 metres of slope foot alluvial and colluvial wash, are the deposits of the debris slope. These deposits differ from the overlying soils. Their permeability is low, <20 mm/hr or less, and both the Plasticity Index (>30) and Liquid Limit (>40) are much higher. They have a significant shrink-swell capacity and sport well-developed desiccation crack polygons. The soil erodibility (K) factor of the USLE is higher at 0.24.

Erosion pins were established at the slope foot, in the upper and lower debris slope, in the subsoil and topsoil both in the gully channel and to either side on its interfluves. Ground loss, change in the exposure of the erosion pins, was monitored semi-annually for five years and total losses converted to sediment yield by multiplying the total ground loss by the bulk density (1.4 g/cm^3). Five-year ground loss averaged 32.5 mm/year on the interfluve in a period when gully depth was 17 mm/year due to infilling. USLE calculation awards the site a mean sediment yield of 1310 t/ha/year but actual measured losses were near 520 t/ha/year (Haigh, 1984).

Representative soil samples were collected for all four zones of the road cut: topsoil, subsoil, debris slope and outwash. These were subject to particle size fractionation by the wet sieve and hydrometer methods. Two samples were collected for each section. The silt-clay fraction of the A-horizon is 22%, the B-horizon 24-26%, and the wash deposits 28-29% but that of the debris slope is 55-72%, so in geotechnical terms, the Darsil (D_{10}) is very fine sand as is the wash deposit, but the debris slope is clay. The textural characteristics of the wash deposit (C), which is derived from the other three, most closely resembles the subsoil. Records of ground loss, or in the case of the debris fan ground advance, were reconstructed in terms of the area of catchment surface occupied by a given material and the percentage of the eroding material in each of the size fractionations (Table 2). This compares the proportions of the particle size fractions contained by the eroded soil in the wash fan with the sediment sources in the catchment. In sediment yield terms, approximately ten times more sediment seems to have been mobilised in the five-year period than was retained in the slope foot accumulation.

Table 2: Comparison of particle-size characteristics of mobilized sediment with the proportion retained by deposition at the road-bank foot

Fraction	*6.0-2.0 mm*	*2.0-0.6 mm*	*0.6-0.4 mm*	*0.4-0.2 mm*	*0.2-0.1 mm*	*0.10-0.06 mm*	*>0.06 mm*
Eroded soil	4.6	0.8	1.1	17.4	22.4	6.7	46.9
Retained by deposition	0.4	0.05	0.04	1.7	3.4	1.3	3.0
Relative retention	9.4%	6.3%	3.7%	9.7%	15.3%	19.7%	6.3%

'Lost Soil' Calculation

The question addressed is how easily can the particle-size characteristics of the 'lost soil', the mobilized sediments, be reconstructed by scaling the soil loss record at each erosion pin by the particle-size parameters of the source? Ideally, it might be hoped that this particle-size distribution would be preserved in the slope-wash/alluvial fan deposits accumulating at the road-bank foot. Normally, one would expect that, since the largest soil particles are least readily mobilized, and since there is little scope for mechanical disintegration in a small catchment, it should be expected that the coarse fragments should dominate the retained slope foot deposits. It is a matter of common observation that while runoff water is dominated by fines, channel floor and fan deposits are dominated by cobbles.

Unfortunately, this is not the case, gravel elements are retained in precisely the same proportion that they are mobilized (Table 2). Further, the pattern of sediment retention is curiously bimodal. Relatively small proportions of coarse and coarse-medium (2-0.4 mm) sized particles are retained and the same is true of particles smaller than 0.06 mm, the silt/clay fraction. On the other hand, a relatively large proportion of particles in the fine to very fine sand size range (0.2-0.06 mm) are found in the wash deposits. Of course, the results may be skewed by the presence of particles that are temporarily in transit.

However, the average retention is about 10% of the amount mobilised and the coarse fraction at 9.4% is almost exactly one tenth. In other words, if you follow the lost soil procedure and calculate sediment yields based on rescaling the particle size classes, the final result is within 10-20% of the recorded total, which is within the range of the sediment yield calculation methodology. So, this study may not have solved the problem, but it does suggest that if you know the proportion of a soil that is in the larger size ranges and can measure the weight of these sized materials in a sediment trap, then it is possible to estimate total sediment production of soil by scaling across each of the sediment classes and subtracting relatively small amount retained in the trap.

Discussion

Orr (1996) has tackled this same problem in the context of forest haul road erosion. His approach also compared soil textures with the source and receiving areas for sediment. He also works from the ratio of coarse to finer materials, using an assumption based on soil textural analysis, that all or some portion of the sediments mobilised was trapped in a 'concentration zone', where they might be measured, and estimates the rest from the depletion of smaller size fractions. However, his case study was of an instance where nearly all the sediment was trapped, unlike the present case.

A tillage study has found that coarse sand proportions tend to remain constant in eroded sediments despite soil aggregation influences (Basic et al., 2002). However, at a larger scale, MacArthur et al. (1990) point out that much, or most, of the sediment that reaches alluvial fans in single events is in fact sourced from the bed and wash load temporarily stored in the channel. Another major source is the channel banks, which may themselves be reworked fluvial or debris flow sediments with a very high percentage of coarse materials (Miller et al., 2004). Both of these factors may seriously distort these calculations.

Conclusion

It is self-evident that there is a direct relationship between soils and sub-soils that suffer erosion and the sediments that are deposited as the result of erosion in different sediment traps. Using the observation that coarse sediments are trapped preferentially, it seems logical to suggest that by scaling the size coarse fraction to that contained in the soils and regolith of the source area, it should be possible to estimate the total amount of sediment mobilised. A small scale case study finds that it is possible to obtain this match in a basin where 90% of all the debris generated moves beyond the slope/gully foot fan. However, further research finds many problems. First, it is difficult to gain an accurate sample for the coarse debris in source sediments because of their variety and the volume needed. Second, erosion is selective and removes low density and platy shaped particles preferentially and smaller size fractions differentially. Finally, the sediment that is deposited in the short term may not represent the source materials directly but originate from the bed and wash load of a channel or atypical alluvial or landslide deposits sourced from a channel bank. However, as Orr (1996) has demonstrated previously, given suitable conditions, a homogenous source area and an assessable relationship between the coarse particle sizes in a deposit and its source area, then this 'Lost Soil' technique may work.

References

Basic, F., Kisic, I., Nestroy, O., Mesic, M. and Butorac, A. (2002). Particle Size Distribution (Texture) of eroded soil material. *Journal of Agronomy and Crop Science*, **188/5:** 311-322.

Bourlier, R. (1976). Soil Survey Interpretation: I Stephenville-Darsil-Windhorst Complex. Oklahoma State University, Department of Agronomy, Stillwater, OK.

Bunte, K. and Abt, S.R. (2001). Sampling Surface and Subsurface Particle-Size Distributions in Wadable Gravel- and Cobble-Bed Streams for Analyses in Sediment Transport, Hydraulics, and Streambed Monitoring. United States Department of Agriculture Forest Service, Rocky Mountain Research Station General Technical Report RMRS-GTR-74; pp 241-283. Available at http://www.fs.fed.us/rm/pubs/rmrs_gtr74/06_chap5.pdf (accessed June 2007).

Goodman, J. and Haigh, M.J. (1981). Slope evolution on abandoned spoil banks in eastern Oklahoma. *Physical Geography*, **2/2:** 160-173.

Haigh, M.J. (1984). Micro-erosion processes and sediment mobilisation in a road-bank gully catchment in central Oklahoma. *In:* Burt, T.P. and Walling, D.E. (eds), Catchment Experiments in Fluvial Geomorphology. Geobooks, Norwich, UK, 247-264.

Haigh, M.J. (1979). Environmental geomorphology of the Landour-Mussoorie area, UP: Preliminary report. *Himalayan Geology*, **9:** 657-668.

Haigh, M.J., Jansky, L. and Hellin, J. (2004). Headwater deforestation: A challenge for environmental management. *Global Environmental Change*, **14/S1:** 51-61.

Lenzi, M.A. and Marchi, L. (2000). Suspended sediment load during floods in a small stream of the Dolomites (northeastern Italy). *Catena*, **39/4:** 267-282.

Loveland, P.J. and Whalley, W.R. (1991). Particle size analysis. *In:* Smith, K.A. and Mullins, C.B. (eds), Soil Analysis: Physical Methods. Marcel Dekker, New York, 271-329.

MacArthur, R.D., Harvey, M.D. and Sing, E. (1990). Estimating sediment delivery and yield on alluvial fans. United States Army Corps of Engineers, Technical Paper 130. Available at http://www.hec.usace.army.mil/publications/TechnicalPapers/TP-130.pdf (accessed June 2007).

Martin, D.A. and Moody, J.A. (2001). The flux and particle size distribution of sediment collected in hillslope traps after a Colorado wildfire. Proceedings of the 7th Federal Interagency Sedimentation Conference, Reno, Nevada, pp. III, 40-47. Available at www.brr.cr.usgs.gov/projects/Burned_Watersheds/Files Martin_Moody_FISC7_2001.pdf (accessed June 2007).

Meijerink, A.M.J. (1974). *Photo Hydrological Reconnaissance Surveys*. Vrije Universitet te Amsterdam, Academisch Proefschrift.

Miller, D.M., Brye, K.R., Morris, T.L., Formica, S.J. and Eps, M.A. van (2004). Estimating bulk density in vertically exposed stoney alluvium using a modified excavation method. *Journal of Environmental Quality*, **33/5:** 1937-1942.

Orr, B. (1996). A rapid method of estimating soil erosion using trapped sediment: A case study from a forest road on the Cumberland Plateau of Tennessee. *Mountain Research and Development*, **16/4:** 419-423.

Walling, D.E. (1988). Measuring sediment yield from river basins. *In:* Lal, R. (ed.), Soil Erosion Research Methods. Ankeny, IA, SWCS/ISSS, 39-74.

Webb, R.H., Griffiths, P.G., Melis, T.S. and Hartley, D.R. (2000). Sediment delivery by ungaged tributaries of the Colorado River in Grand Canyon. U.S. Geological Survey Water-Resources Investigations Report 00-4055, 1-67.

Wood, P.R. and Burton, L.C. (1968). Groundwater resources in Cleveland and Oklahoma Counties, Oklahoma. *Oklahoma Geological Survey, Circular*, **71:** 1-75.

22

Socio-Economic and Environmental Impacts of Erosion and Sedimentation in Sudan

Seifeldin H. Abdalla
Ministry of Irrigation and Water Resources
Khartoum, P.O. Box 878, Sudan

Introduction

Severe soil erosion is experienced in the watersheds of the Blue Nile and Atbara rivers originating from the Ethiopian highlands. The resulted sediment causes sedimentation of dams' reservoirs, power houses, inlet channels of pumping stations and irrigation canals. Storage capacities of reservoirs are reduced to 50% due to sedimentation. The high silt content during the flood periods causes frequent blockage of turbine inlets of the major hydropower station at Roseries dam resulting in severe power shortages. Inlet channels of the major irrigation schemes in the Northern region are now severely affected by sedimentation and morphological changes of the river Nile. Inlets are clogged with sediment and are now inoperative. Others will follow affected by the steady propagation of the problem. In irrigation canals, sedimentation affects directly the operation and maintenance of the irrigation systems, draining annual maintenance budgets and causing severe water shortages and crop damage. Bank erosion in Northern Sudan is causing tremendous problems to local people, reducing their already small cultivable lands. In fact it is not only a problem of loosing lands, crops and power by erosion and sedimentation, but it has significant social, economic and environmental implications as well.

In this article, several areas in Sudan affected by sedimentation and erosion are surveyed (reservoirs, inlet channels, power stations, irrigation canals, river banks, etc.), focussing on their economic, social and environmental impacts. The article also reviews the most significant efforts exerted to mitigate such impacts, and suggests the best methods and means to alleviate them.

Erosion in Watersheds

Most of the sediment coming to Sudan originates in the Ethiopian highlands. The Blue Nile and Atbara rivers transport very high sediment concentrations in the summer flood season. Eighty percent of the annual discharge of the two rivers occurs between July and October. During this period the two rivers transport very high sediment concentrations reaching 2.6%, by weight recorded during the historic flood of 1988 (Demas, 1988). The sediment concentration vary from year to year, but the average peak is about 1% by weight at the end of July and it reduces to few hundreds ppm by the end of October (Hamid, 2002). The transported sediment consist of significant quantities of silt and clay (D<63 microns) known as wash load which can easily be transported in suspension.

Sediment Entering the Irrigation Systems

Table 1 shows the mass of sediment entering the Gezira irrigation system (the largest irrigation system in Sudan, 900,000 ha) in thousands of tons per month taken as an example of the sediment entering the national irrigation schemes in Sudan which are irrigated from the Blue Nile and Atbara rivers. The sediment entering at the rest of the year is almost negligible compared to the amounts shown in Table 1.

Table 1: The mass of sediment entering the Gezira main canals (thousands of tons per month)

Month/Year	*1981*	*1982*	*1983*	*1984*	*1987*	*1988*
July	327	705	367	905	910	2129
August	2327	2780	4625	2259	1663	2067
September	1002	949	2744	858	359	1115
October	341	190	464	205	175	379
November	85	173	179	113	158	229
Total	4120	4697	8379	4390	3277	5919

Source: Sir M MacDonald and Partners Limited, 1992.

Sedimentation in Irrigation Canals

The distribution of sediment in the Gezira system was studied by the Hydraulic Research Station (HRS), Ministry of Irrigation (MOI), at Wadmadani, Sudan in collaboration with Hydraulics Research Limited (HRL), Wallingford, England. Sediment load evaluations have been made along the main canal, along selected majors and selected minors (HRL, 1991). The principal conclusions drawn from a preliminary analysis of the data collected in 1988 are:

- Of the sediment arriving at Kilometre 57, the head of the system, approximately 14% settled in the main canals, 22% settled in the major canals, 33% in the minors and 31% was passed to the fields.
- In most reaches, the major part of the sediment deposition occurred in the last two weeks of August, the time when sediment concentrations entering the system were at their highest.
- In minors very little scour is observed when sediment concentrations are reduced, or when discharge is increased. For cohesive sediments, bed level rise when sediment is deposited, but do not fall again when the flow is increased or when incoming sediment concentrations decrease. Cohesive sediments consolidate after deposition so that re-suspension would require very large shear forces.

Sedimentation in Fields

There is a variation in the pattern of deposition in the fields with locations along the major, along the minor, and along Abu XX (small tertiary canal irrigating the field). Francis and Hinton (1988) reported an average rise in the ground level at the heads of the fields of 0.30 m since the mid 1920s, with a larger rise at the head of a minor than at the tail. This implies that even if the problem of sediment deposition is solved, high water levels in minors will be required in the future to pass design discharges through the field outlet pipes to Abu XXs (Abdalla, 1992).

Sedimentation Estimates from MOI Records

For the Gezira scheme, MOI figures for estimated and extracted sediment volumes for some selected years are shown in Table 2. Approximately half of the sediment entering the system is removed. If only 31% is being deposited in the fields, the difference is accumulating in the canal system. From Table 2, it is clear that estimates for annual desilting needed varies from 6.5 to 12.5 million m^3. Last years records of the quantities of sediment removed from the Gezira scheme show an enormous increase in the rates of soil erosion in the watershed enhanced by the effect of successive droughts in the Ethiopian Plateau.

Table 2: Selected MOI sediments records for Gezira scheme (million cubic metres)

Year	*1982*	*1983*	*1984*	*1985*	*1988*	*2000*	*2001*	*2003*	*2004*
Estimated inflow	10.0	9.9	8.1	6.5	7.0	10.0	12.50	9.20	10.60
Removed from canals	5.4	4.4	7.0	5.1	6.5	7.92	11.50	12.16	11.81

Source: MOI Annual Reports, 1981 to 2004.

The problem of sediment deposition in the canalization systems of the irrigated schemes poses a great threat to the national economy and a serious problem to the farmers, and believed to be the main factor causing water shortage, interrupting water supplies, and increasing incidents of canal bank overtopping due to high water levels, creating a conducive environment for harbouring weeds, and hence lowering the production per unit hectare. The sediment is usually removed mechanically by using hydraulic excavators and draglines, to restore the original design sections of the canals (Abdalla, 1992). The cost of canal clearance has become a major item in the operation budget of these systems due to the escalating price in the cost of the unit silt removal (MOI Annual Report, 2004).

Effects of Sedimentation on Dams' Reservoirs

There are at present three dams on the Blue Nile and Atbara Rivers: Roseires (1966) and Sennar (1925) are situated on the Blue Nile, and Elgirba (1964) on Atbara River. The dams were constructed to provide storage for irrigation in the low water season and for hydropower generation. Although the dams have deep sluices set at the lowest possible level in the main river channel and the proper operation schedules were followed during the flood periods but still reservoirs storage capacities were significantly reduced as a result of sedimentation. Heavy siltation have taken place during the first years of operation (Ahmed and Hamad, 1986). In recent years a progressive reduction in sedimentation rate is reflected in the recent surveys which indicate that the reservoirs reached a state of equilibrium. The consequences of the unexpected amounts of sedimentation of the above reservoirs are manifested in the loss of agricultural lands associated with the construction of these dams. For example, the New Halfa agricultural scheme (180,000 ha) for which Elgirba dam was constructed has lost more than one third of its cropped area due to the depletion of the storage capacity of the Elgirba reservoir in addition to the curtailment of the new development of the region (MOI Annual Report, 2004). Flushing operations have been carried out regularly for Elgirba reservoir since 1971 in order to reduce the rate of siltation. Table 3 shows the remaining capacities of the reservoirs.

Table 3: Design capacities compared to current capacities of silted reservoirs

Reservoir	*Design capacity (million m^3)*	*Current capacity (million m^3)*	*Percent loss of capacity*
Roseires	3400	2000	41
Sennar	930	400	57
Elgirba	1300	600	54
Approximate average			50.7

Source: MOI Annual Report, 2003.

In Roseires dam, the accumulation of silt in front of the power house intakes with its floating debris during the flood season restrict to a large extent the power generation and resulted in severe power shortage during the flood season.

In general, as a result of the reduction in reservoirs storage, the availability of water for the existing agricultural schemes is in doubt. It is estimated that to maintain an acceptable security of supply for the existing schemes, hydropower generation, and minimum low river flows in the Blue Nile, reservoir storage of Roseires must be increased. With the reservoir raised 10 m, storage would be sufficient for sustaining the existing schemes and developing new ones.

The Effect of River Morphological Changes

The river Nile is a typical alluvial river. In northern part of Sudan it passes through alluvial plains surrounded by an extended desert. Its banks and bed continuously experience a series of changes. Complexity of changes is manifested in the form of meanders, narrow and wide channels, shallow and deep pools, flat and steep banks and slopes, rocky and silty beds, etc. and these are never permanent (Adam and Abdalla, 1988). These changes ultimately affect the cultivable areas. It is known that the cultivable area in the northern region depends mainly on the Nile waters in irrigation.

Water started to be inaccessible in different parts of the region adding a burden on the hardships already existing in the area. This is attributed to the dilemma of sedimentation, river widening and bank erosion. These problems have paramount social, environmental and economic implications. However, bank erosion in the northern region receives much of the attention but sedimentation and recession caused by river widening are causing no less damage than the former (Adam and Abdalla, 1989). The impact of bank erosion is much greater, but abandonment of agricultural schemes results in much devastation. In the next part the magnitude of recession and sedimentation problems are highlighted including causes and measures being taken by Government and local people (Table 4).

Sedimentation of Inlet Channels of Pumping Stations

The total area of irrigated agriculture in the northern region is around 130,000 ha and expected to rise to 250,000 ha in the next ten years (Abdalla, 2002). Two thirds of this area is cooperative and private schemes and one third is governmental schemes. There are 114 cooperative schemes covering an area of 40,000 ha. The area of the scheme ranges between 40 ha for small schemes and 800 ha for large ones. All of these schemes were irrigated from the Nile through pumps driven by diesel engines. A survey of 25 of the cooperative schemes (Abdalla, 2004) revealed that most of these schemes are now in poor conditions suffering from sediment deposition in front of the intakes (inlet channels) of the pumps. Many of the pump stations were clogged and are now inoperative; others will follow affected by the steady propagation of the problem.

Excessive sediment accumulation problem was first encountered during 1975-1980. In spite of the steady development of the problem, no radical technical solution or protective measures were sought by the farmers or the authorities. The situation is now critical and represents the major constraint towards full utilization of these highly potential schemes, the productivity of which is incomparable to any other scheme. The production records buttress this fact, as it shows that:

- The wheat productivity of these schemes is 4.2 times higher than the productivity of the Gezira and the other main national schemes.
- The cropping intensity in the abundance of irrigation water supply is more than 150% compared to 70% in the other national schemes.
- Crops quality is better than other schemes.

Owing to these reasons, it is possible to state that every effort to solve the sedimentation problem is worthwhile.

This sedimentation problem of the intakes of the irrigation schemes is part of the overall intricate and complex change of river bed morphology which differs in many ways from the relatively simple siltation problem:

- Morphological changes are a global phenomenon affecting the whole river reach, whereas siltation is a local, limited phenomenon.
- The mechanism of phenomenal initiation, characteristics of development and approach for solution are extremely different for each phenomenon.
- Seasonal silt clearance operations are effective for general practice solution for inlet channel siltation problem, which fails completely to solve the other problem except at early stages of its progress and at very high costs.
- Siltation is a recurrent phenomenon, confined to a certain location, whereas morphological changes are a multiple dimensional, steadily progressing problem.
- Morphological changes phenomenon is inseparable from other inter-related phenomena: i.e. bank erosion, formation and disappearance of river islands, river braiding etc.

The history of this phenomenon does not go back to more than the last three decades when those changes were noticed in some schemes. The problem was aggravated in the last years due to increasing extraction of irrigation water upland together with natural decreasing river inflows. Presently, the problem development is awful, and seriously threatens beneficial utilization of this

Table 4: Selected areas of bank erosion and recession in Sudan, surveyed by HRS in the period 1989-1999

No	*River reach*	*Location affected*	*Type of problem*	*Level*	*Area endangered*	Solution applied
1.	Blue Nile (Wad Medani-Khartoum)	Wad Medani town	Bank Erosion	L	Wad Medani Town	Retaining wall
		Gunied pumping station	Recession	M	Gunied sugar scheme	N.A.
		El Masseed Village	Bank Erosion	S	EL Masseed village	N.A.
		El Bashagra East Village	Bank Erosion	S	Bashgra east village	N.A.
		Presidential Palace	Bank Erosion			Retaining wall
		Tuti Island	Bank Erosion			Pitching
2.	Main Nile (Khartoum Area)	Shambat	Bank Erosion	M	Shambat bridge and Agric. Lands	N.A
3.	Nile State reach	Gundato pumping station	Recession	S	Main line (Khartoum/ Port Sudan)	Floating pumps
		Kelli pumping station	Recession	M		Annual dredging
		Pipe line/Atbara river	Bank erosion	S		Groynes and pitching
4.	Northern State reach	Nuri	Bank erosion	S	Pumping station and 2100 ha	Groynes and pitching
		Merwei east	Bank erosion	S		Pitching
		Abu Doum (Photo 3)	Recession	S	70% of Agric. Land lost	Annual dredging
		Tangssi (Photo 1)	Bank erosion	M	1260 ha abandoned	N.A.
		Azzuma	Bank erosion	S	10% of Agric. Land	Pitching
		Gurair Agric. Scheme	Recession	M	Pumping station and 378 ha	N.A.
		Fageerankutti	Bank erosion	L	1050 ha suffering	N.A.
		Ganatti	Bank erosion	M	Agric. Lands	N.A.
		Elkulud	Bank erosion	M	Agric. Lands	N.A.
		Kurmakole	Bank erosion	M	Agric. Lands	N.A.

(*Contd.*)

Table 4 (*Contd.*)

	El Dabba	Bank erosion	M	Agric. Lands	N.A.
	Ghaba Agric. Scheme	Bank erosion	S	Agric. Lands	N.A.
	Mussawi Island	Recession	L	Agric. Lands	Groynes
	Elbasa	Recession	M	84 ha abandoned	N.A.
	Elhigeir	Recession	M	660 ha abandoned	N.A.
	Elghuraiba	Recession	S	105 ha abandoned	N.A.
	El arak	Recession	M	588 ha abandoned	N.A.
	El golid gubli (Photo 2)	Recession	S	140 ha abandoned	N.A.
	Urbi	Bank erosion	L	483 ha abandoned	N.A.
	Sheikh Sharief	Bank erosion	M	35% of Agric. Land lost	N.A.
	Labab & Grada Islands	Bank erosion	L	Agric. Lands	N.A.
	Magasir Island	Bank erosion	S	Agric. Lands	N.A.
	Burgaig Agric. Scheme	Bank erosion	S	Buildings and Agric. Lands	N.A.
	Binna Island	Bank erosion	S	60% of Agric. Land lost	N.A.
	Artigasha Island	Bank erosion	S	Agric. Lands	N.A.
	Badeen Island	Bank erosion	M	Agric. Lands	N.A.
	Karma el nuzul	Bank erosion	S	Karma Elnuzul village	Pitching

Note: N.A.: Not available; S: Severe; M: Moderate; L: Light.

very high productive region. The irrigated areas have continuously decreased to small percentages of the total irrigable areas, the figures range between 20 and 90%. Moreover, even these small proportions are irrigated by means of very small pump units distributed in the fields to extract ground water to feed small plots of 1.2 to 2.1 ha for each unit. This irrigation method is extremely inefficient and suffers the following drawbacks:

(a) Only limited number of the farmers has the interest and financial capability to buy and run such units (high running cost). Also the high consumption of spare parts and fuel created non-availability problems.
(b) Use of ground water deprives the land from rich and renewable natural fertilizer carried by the river.
(c) High silt concentration of the irrigation water reduces infiltration rate, and hence increases the irrigation application efficiency, decreasing the total crop water requirement and the irrigation frequency.
(d) Discharge of these small pumps is very limited, only 0.21 to 0.30 ha could be irrigated every day; this requires intensive attendance from the farmers' side.
(e) Extended practicing of small units' irrigation may enforce division of co-operative schemes into small actions and create difficulties in future if regrouping and modernization plans are to be implemented.

River Bank Erosion and Recession

The Nile reach in Sudan can be divided into four reaches. The first reach is between Khartoum and the fifth cataract. Second reach is from the fifth to the fourth cataract (Hamadab). Third reach is between the third cataract and fourth cataract. The last reach is from the fourth cataract to Aswan High Dam Lake. Of these four reaches only the first and the third are reported to have noticeable bank erosion. The second and the fourth reaches are of quite stable banks. However, the bank erosion in the first reach is not severe as that of the third reach.

The river Nile downstream of Hamadab area, undergoes severe bank erosion at many different locations as far as Dongola, 470 km downstream. Because of the adverse effects of the river bank erosion on the economy and living standards of the people in that area, various protection measures such as groins and revetments were constructed during the last 60 years in areas of erosion which shifts frequently. In the period 1979-1989 the Hydraulic Research Station (HRS) of the MOI carried out intensive field programmes to examine the erosion and recession problems along the Nile and its tributaries (Adam and Abdalla, 1988). In these studies 37 severely affected locations were surveyed using an echo-sounder (Table 4). Point velocities were also taken in different cross-sections in the affected locations, at a variety of depths and locations, using current meters. Part of this information was used to construct a rigid boundary, physical model to study the flow pattern of the reach that extends 20 to 60 km from the Hamadab region which suffered from bank erosion resulting in great damage to private and public properties (Manohar and Adam, 1982). The study concluded that the main source of erosion is the excessive kinetic energy manifested by excessive energy gradient and stream flow during the flood time. Also the study indicated that the relative increase in energy was less for the highest floods, so that lesser floods caused more damage by erosion than the highest floods in that reach.

Examples of erosion and recession problems are illustrated in Table 4 and Figs 1, 2 and 3. Figure 1 shows typical bank erosion at Tangassi, where 6 km were affected, 1500 palm trees and 10% of very fertile land were lost. Figure 2 shows sedimentation at Elgolid gubli where 483 ha of very fertile land was abandoned because water was recessed for more than 150 m from the pump intakes. Hopelessly, the local farmers purchased a dragline in order to combat sedimentation, but their efforts were gone astray. Figure 3 reflects recession problems at Abu Doum where 13,000 inhabitants were affected and 1260 ha of fertile land was abandoned. The local people tried in vain to excavate inlet channels to alleviate the problem. It is worth noting that the agricultural area could be extended largely for the welfare of the inhabitants if the accessibility of water is secured.

Figure 1: Erosion in the vicinity of the pump station at Tangassi (Table 4).

Figure 2: Sedimentation in front of the pump house at Elgolid gubli scheme (Table 4).

Figure 3: Recession at Abu Doum scheme pumping site (Table 4).

Conclusions and Recommendations

The Blue Nile and its tributaries and Atbara river, originating from the Ethiopian plateau, are the main source of sediment in the canalization systems, dams' reservoirs, and inlet channels of the pumping schemes. One of the measures to alleviate the problem is to stabilize the upper watershed areas, thus minimizing soil erosion and land degradation, with ultimate result of making the highlands unable to support sediment. It is worth noting that watershed management is now a major component of Nile Basin Initiative, Subsidiary Action Programs, and the FRIEND/NILE project. Coordination of efforts, cooperation, synergy and linkages are very much required at this stage to tackle the problem from a regional perspective.

The problem of sedimentation in the Sudanese irrigated systems has been a serious one, and represents the major cause of water shortage in many parts of these systems, reducing the production and affecting the livelihood of the inhabitants of the scheme. One of the most effective control measures is to reduce flows entering the canal system in late July and early August. Closing the main canal over this period would reduce the sediment input by at least 50%.

Sedimentation of the minor canals is proved to be causing the main problems in the irrigation schemes. Provision of sediment storage in minor heads by over-digging would provide storage for the sediment, so that deposition can be accommodated without increasing the water level at the minor head. To minimize sediment deposition in majors, the idea of constructing sediment traps at the heads would be useful. Also slopes should be allowed to be steepened, to the extent that the inlet gate characteristics and upstream bank levels allow increased water depths.

Basins could be designed to store at least one year's sediment deposition without affecting the flow to the major, and would be desilted annually. The advantage of this option is that desilting would be concentrated at a few locations. The disadvantages include the need to dispose off the dredged sediment, the loss of some agricultural land and the need to realign some upstream minors.

The consequences of sedimentation of dams' reservoirs on the economy were far reaching. Agricultural development in Sudan by necessity depends on large storage facilities to augment meagre dry season flows. Sedimentation should, therefore, be considered as an important parameter in the design of future works. Options of sediment control measures are limited to the watershed management, appropriate operating schedule during the flood period, flushing operation of reservoirs, and the annual dredging in front of the inlets of the power turbines.

The main cause of the recession and sedimentation problems facing the inlet channels of the pumping schemes is the small flows and low levels occurring during the summer period. If large storage is secured, then it is possible to regulate the flow in order to obtain desirable levels. Merawi dam, which is now under construction at Hamadab, the proposed Roseires heightening and Jonglei canal are likely to fulfill that objective. The effect of Merawi dam would be more noticeable due to its site location in the affected region. Raising the water levels using cross weirs would also be a feasible solution. This would result in raising the ground water levels too, thus farmers could extract irrigation water from wells with lower cost. In addition, the use of mobile pumping stations proved to be very successful. However, care should be taken to minimize pipe friction losses and to keep good watch on maintenance and level variations.

The total length of severely eroded banks is approximately 40 km, resulting in high economic and environmental impacts. These impacts are manifested in loss of fertile agricultural lands, crops and high value perennial trees and dislocation of families and villages. Protection using revetments was found to be more stable and sustainable but because of the high cost involved, small reaches are now being protected.

References

Abdalla, Seifeldin (1992). Water management in the Gezira irrigation scheme, Sudan. Report submitted for Utah State University, Logan, Utah, USA.

Abdalla, Seifeldin (2002). Long term strategy (2002-2027) of irrigated agriculture in Sudan: The Nile section. MOI Report, Khartoum, Sudan.

Abdalla, Seifeldin (2004). Sediment deposition problem in Northern Province co-operative schemes: Short term solution. MOI Report, Khartoum, Sudan.

Adam, A. and Abdalla, Seifeldin (1988). Feasibility study about bank erosion in the Northern Province. HRS report, Wadmedani, Sudan.

Adam, A. and Abdalla, Seifeldin (1989). Implications of river widening in view of bank erosion in the Northern Province. HRS Report, Wadmedani, Sudan.

Ahmed, T. and Hamad, O. (1986). Sediment transport in relation to reservoirs. Proceedings of the International Conference on Water Resources Needs and Planning in Drought Prone Areas. Khartoum, Sudan. Part one, 281-295.

Demas (1988). Comparison of rate of siltation in Night Storage Irrigation canals vs Continuous Irrigation canals. MOI, Wadmedani, Sudan.

Francis, M. and Hinton, R. (1988). Minor canal management in Gezira: Field investigations on selected minor canals. HRL, Report OD 106, England.

Hamid, S. (2002). The problem of sedimentation in Nile basin irrigation systems: The case of the Sudanese irrigation systems. Proceedings of the 9th Nile 2002 conference, Nairobi, Kenya.

HRL (1991). Research for rehabilitation: Study of reliability of water supply to minor canals. Vol. 1, Hydraulic Research Limited, England.

Manohar, M. and Adam, K. (1982). Stability analysis for flow boundary at Nuri-Kareima stretch. HRS Report, Wadmedani, Sudan.

Ministry of Irrigation and Water Resources (MOI). 1981 to 2004 Annual Reports. Khartoum, Sudan.

Sir M MacDonald and Partners (1992). National irrigation rehabilitation program for irrigation schemes in Sudan. MOI, Wadmedani, Sudan.

Index

GPSR Compliance

The European Union's (EU) General Product Safety Regulation (GPSR) is a set of rules that requires consumer products to be safe and our obligations to ensure this.

If you have any concerns about our products, you can contact us on ProductSafety@springernature.com

In case Publisher is established outside the EU, the EU authorized representative is:

Springer Nature Customer Service Center GmbH
Europaplatz 3
69115 Heidelberg, Germany

Batch number: 10823835

Printed by Printforce, the Netherlands